The New Renaissance

The New Renaissance
Computers and the Next Level of Civilization

Douglas S. Robertson

New York   Oxford
Oxford University Press
1998

Oxford University Press

Oxford   New York
Athens   Auckland   Bangkok   Bogotá   Buenos Aires
Calcutta   Cape Town   Chennai   Dar es Salaam   Delhi
Florence   Hong Kong   Istanbul   Karachi   Kuala Lumpur   Madrid
Melbourne   Mexico City   Mumbai   Nairobi   Paris   São Paulo
Singapore   Taipei   Tokyo   Toronto   Warsaw

and associated companies in
Berlin   Ibadan

Published by Oxford University Press, Inc.
198 Madison Avenue, New York, New York 10016

Oxford is a registered trademark of Oxford University Press

Library of Congress Cataloging-in-Publication Data
Robertson, Douglas S.
The new renaissance:
computers and the next level of civilization /
by Douglas S. Robertson.
p. cm. Includes bibliographical references (p.) and index.
ISBN 0-19-512189-9
1. Computers and civilization.
QA76.9.C66R618 1998 303.48 34-dc21 97-31239

Chapter 1 is revised from Robertson, D.S., 1990.
"The Information Revolution."
Communication Research 17, no. 2: 235-54.
Copyright © 1990 by Sage Publications, Inc.
Reprinted by permission.

Typeset in OCR and Minion
Design and composition by Adam B. Bohannon

1 2 3 4 5 6 7 8 9
Printed in the United States of America
on acid-free paper

To the memory of my mother, Margaret Ellen Clayton Robertson—
who made everything possible.

To my wife, Joan—
who believed in me and read every draft.

And to my computer-literate daughter, Carolyn—
the future is hers.

Contents

Introduction 3

1 The Information Revolution 8

2 "Theories of Everything"
and the New Copernican Revolution 37

3 The Computer Revolution
in Science and Mathematics 57

4 Uncomputable Numbers 71

5 The Computer Revolution
in Education 93

6 Language in the Computer Age 114

7 Decimal Delenda Est 131

8 The Computer Revolution
in the Arts 143

9 The Impact of Computers
on Everyday Life 160

10 On Growth 171

Conclusion 179

References 189

Index 193

The New Renaissance

Introduction

The invention of the computer is one of the pivotal events in the history of civilization. As far back as the early 1960s, computer technology was already having an enormous impact on science, mathematics, engineering, business, and commerce. But today the impact of computer technology reaches far beyond these areas. It touches virtually every field of human activity, and it has transformed many of them beyond recognition.

This book examines the computer revolution from a historical standpoint. It develops new perspectives by exploring the effects of similar revolutions in the past. At least three earlier technological innovations transformed civilization in ways that provide important historical parallels to the computer revolution.

My interest in this subject started with a fascination for the mathematical revolutions that lie at the root of computer technology. I first encountered some of these startling mathematical innovations when a gifted teacher in grade school introduced me to George Gamow's remarkable book *One, Two, Three,...Infinity* (1961).

In this book Gamow described Cantor's discovery that there is more than one infinity. He went on to explain why it makes sense to speak of larger and smaller values of infinity. Like many people, I first reacted to these ideas with incredulity and disbelief. But my next reaction was astonishment when I realized that Cantor's work, for all its elegance and power, was stunningly simple. It was simple enough that a grade-school student could follow it and understand the ideas and arguments without serious difficulty.

I began to follow the further development of Cantor's ideas by Hilbert, Russell, Whitehead, and especially by Gödel and Turing. It seemed incredible that questions such as "Is mathematics consistent?" and "How many axioms are required to span all of mathematics?" could even be posed, let alone answered. I was even more amazed that the answers turned out to be so counterintuitive. Newman's lucid discussion of Gödel's theorem (along with the rest of the fascinating series he edited, titled *The World of Mathematics* [1956]) had a particularly strong impact on my thinking.

I had stumbled onto some of the most interesting and important discoveries in the history of mathematics. Gödel's theorem in particular seemed to me to be the essence of pure mathematics in that it was of immense theoretical and philosophical importance but at the same time was so abstract that it appeared to have little practical importance, even for mathematicians. Indeed, I have heard several mathematicians state that Gödel's theorem has no impact at all on their work, and little effect on the general practice of mathematics.

The practical importance of Gödel's work did not become apparent to me until I arrived at MIT to do graduate work in planetary physics in the late 1960s. There I was profoundly shocked to discover that Gödel's theorem was an essential part of the curriculum, not in mathematics but in electrical engineering. This book is largely the outgrowth of that shock.

What was so extraordinary was that Gödel's investigations into the deepest roots of pure, abstract mathematics, the nature of mathematical truth itself, provided at the same time the theoretical underpinnings for perhaps the most important *practical* aspect of applied mathematics in all history, the electronic computer. In investigating the fundamental nature and limitations of mathematics, Gödel and Turing probed the basic nature of what a mathematical operation really does, and Turing was the first to realize that every operation in mathematics could be performed by a simple machine.

It seemed clear that since the practical importance of this mathematical revolution more than matched its theoretical and philosophical importance, this must be a truly world-shaking occurrence. And it is. Civilization is being changed beyond recognition by the computer revolution.

I began to investigate the practical effects of the computer revolution, the implications of a major change in the way that civilization produces, distributes, and processes information, and I soon found a set of fascinating historical parallels. The first chapter of this book describes these parallels in some detail. It explores the similarity between the invention of the computer and the invention of the printing press, in addition to parallels with the earlier inventions of writing and language.

The next three chapters describe the mathematical revolution that has long fascinated me, but examined now in the context of the parallels to earlier mathematical and scientific revolutions. Chapter 2 examines the correspondence with the Copernican revolution. Chapter 3 describes the parallel with the scientific revolution of the seventeenth century that centered on the invention of the calculus and Newtonian dynamics.

Chapter 4 is the most mathematical chapter in the book, and it is the only one to contain equations and sketches of proofs. Though these may not be to everyone's taste, they get to the heart of the questions that I was investigating. And the chapter is not as difficult as it might seem: The equations are largely ornamentation, and the proofs are confined to boxes at the end of the chapter. The concept of an uncomputable number was so fascinating that I was unable to resist taking the opportunity to discuss it.

The revolutions sparked by computer theory and technology as described in chapters 1 through 4 have brought about changes that are so vast and complete that they represent nothing less than the dawn of the next level of civilization. The generation that is alive today has an opportunity to design the next civilization. This is not something that is given to every generation.

The remaining chapters of the book are less theoretical. They explore what I think will be some of the important practical aspects of the new level of civilization that will be created by the computer revolution. Chapter 5 looks at the impact of computers on the education process. Chapters 6 and 7 examine the impact on the English language and on ordinary arithmetic. Chapter 8 looks at some of the revolutionary possibilities that computer technology provides for the arts. Chapter 9 looks at the impact of computers on our personal lives.

Chapter 10 is the only one that has little to do with computers. Rather, it explores some of the problems associated with the future of civilization. The key problem faced by any civilization is posed by the difficulties of growth, along with the attendant problems of pollution, depletion of resources, and so on. This chapter looks at the mathematics of growth, and shows how narrowly our choices in this matter are constrained by the basic mathematics.

The conclusion examines some of the key features that I hope might be found in the new civilization.

No single writer or book could possibly cover all aspects of the monumental changes involved in the dawning of a new level of civilization, and I have made no attempt at a comprehensive treatment. Rather, this is an exploration of a few of the more compelling effects of the computer revolution on key components of modern civilization. Some of my forecasts of the future of civilization are intended to be outrageous for the simple reason that the future is certain to be outrageous by any standards of today. Projections that are merely reasonable are certain to be wrong. Yet the primary focus of this book is on parallels rather than on projections, with a secondary focus on the extraordinary and mind-bending discoveries that accompanied the computer revolution, discoveries that have fundamentally changed the way we understand our universe. These discoveries have made the revolution that was launched by Copernicus appear modest and tame.

History has never seen a revolution on the scale of the one that is being triggered by computer technology. The closest historical parallels, the revolutions of the Renaissance, occurred on a far smaller scale than the ones being touched off today by computer technology. And just as the Renaissance marked the beginning of modern civilization, this new revolution will mark the beginning of the next level of civilization.

1

The Information Revolution

This has often been called the age of information, and for once a popular label is correct. Never before has information been produced at such a staggering rate, and the rate is increasing rapidly. This explosion of information was spawned by the extraordinary development of modern computer technology, and it has already revolutionized manufacturing and business, science and technology, schools, government, even some of our homes. But the changes have barely begun: The production and use of computers is still growing explosively.

To understand the significance of the information revolution that has been touched off by computer technology we need to investigate its historical parallels. A study of the effects of comparable inventions in the past can provide important insights into the revolutionary implications of computer technology today.

At first it might seem that no such parallels exist, that the computer revolution is unique in history. But it is not unique. At least three inventions in the past—language, writing, and printing—had effects that were very similar to today's computer revolution. Each of these inventions decreased the effort and cost required to produce, store, and distribute information, thereby causing an information explosion very similar to the one being created today by computer technology.

To understand the importance of these three inventions we need to place them in their proper historical contexts: Each is closely associated with the beginning of a fundamentally new form of human society. The invention of language is associated with the very beginning of the human race, the invention of writing with the beginning of civilization, and the invention of printing with the beginning of modern civilization.

These associations are highly suggestive. The most important dividing points in the history of civilization were each accompanied by an invention that caused an information explosion. This suggests a possible cause-and-effect relationship—that information explosions may have caused these transformations of civilization. If we can prove that such a relationship exists, the implications for civilization today would be profound: The computer could be the invention that will change civilization to a degree not seen since the Renaissance, the time of the last great revolution in information handling.

To show that such a causal relationship exists we need to show that civilizations generally are information-limited. In other words, we need to show that a limit on the production of information impeded progress in the period that preceded each information explosion. To do this we must study in detail the limits that are placed on a society by restrictions on the production of information.

The fundamental reason that civilization is limited by information is quite simple: Civilization *is* information. Most of the factors that characterize a civilization—its ethics and laws, its technology, its philosophy and religion, its literature and art—are forms of information. And civilizations are generally limited more by lack of information than by lack of physical resources. Classical civilization, for example, possessed all of the resources needed to create Morse's telegraph and Edison's phonograph. Ramses and Pericles failed to build telegraphs and phonographs only because they lacked the necessary information.

But information limitations have broader implications, beyond the simple lack of specific items of information. Information limits are quantitative as well as qualitative, and their quantitative study can lend broad insights into the patterns that mark the development of civilizations. Just as a quantitative limit on the production of steel restricts the output of automobiles and ships, a limit on the production of information restricts the number of things that a civilization knows how to do. A quantitative information limit restricts not only science and technology, but also such things as literature, politics, art, music, architecture, religion, law, ethics, history, and philosophy. Indeed, there is hardly an aspect of civilization that is not directly restricted by limits on information production. To understand these restrictions in detail we must quantify the information requirements of key elements of various civilizations and compare those requirements to the actual capacity of civilizations to produce information.

If information limits are so important, we might well wonder why they have not already been studied in some detail. There are probably two basic reasons. First, the very concept of information as a quantifiable substance is a very recent development. The quantitative theory of information was one of the great triumphs of twentieth-century science, developed by scientists and engineers who were trying to optimize radio and telephone communication. The theory is so new that even in the physical sciences many researchers are only beginning to understand its full implications for their fields. We should not be surprised that its implications for history and sociology have barely begun to be probed.

The second reason that information explosions have not generally received the attention they warrant is more subtle: No two of them are studied by the same scientific discipline. The first one is studied by anthropologists and linguists, the second one by archaeologists, the third by historians. It is difficult to recognize a recurring pattern from the study of a single instance, and few professionals have the breadth of knowledge necessary for proficiency in all the requisite disciplines. Thus each of these three inventions has generally been studied in isolation, and the recurring pattern of information explosions has remained relatively unexplored.

The vital importance of information to human civilization has not gone completely unnoticed, however. In the 1930s the French historian Henri Berr pointed out the epochal significance of the inventions of language, writing, and printing for the development of civilization (1934). Berr, unfortunately, seems to have been writing a little ahead of his time. The intimate connection among these inventions through their quantitative effect on the production of information could not have been completely understood prior to Shannon's development of a quantitative theory of information in the 1940s. And, more importantly, Berr had no way of knowing that the computer, the next critical invention that would have an explosive effect on information production, was only a decade away.

It might seem extreme to argue that the first of these critical inventions, the invention of language, was the only factor or even the principal factor that distinguished early humans from their prehuman antecedents. Defining the exact point of origin of the human race in the fossil record is somewhat arbitrary. Nevertheless the invention of language is a very good choice for the point that marks this critical juncture. Gould called language "the most widely cited common

denominator and distinguishing factor of humanity"
(1993, 321). Falk, an anthropologist at the State University of New York, puts it this way:

> Various skills, including hunting, tool production, warfare and language have been advocated at one time or another as responsible for human brain evolution. Apes occasionally do all but one of these activities. They sometimes hunt, make tools, and fight, but they never engage in spoken language as we know it. If we wish to identify one prime mover of human brain evolution,…it is language…. Human technology and social achievements required conscious thought, which is, and probably was, dependent on language. In other words, until they acquired language, our early ancestors may not have been truly human. (1984, 38)

The connection between the invention of writing and the dawn of civilization is even clearer. In the third millennium B.C. civilizations emerged almost simultaneously in four so-called "cradles of civilization": Egypt, Mesopotamia, the Indus Valley, and China. The single factor that distinguished these four cultures from their neighbors was the invention of writing. Other factors, such as metallurgy, commerce, and urbanization, were possessed in some degree by other cultures. Writing alone distinguishes civilization. As Gelb of the University of Chicago writes:

> [M]any other great men—among them Carlyle, Kant, Mirabeau, and Renan—…believed that the invention of writing formed the real beginning of civilization. These opinions are well supported by the statement so frequently quoted in anthropology: As language distin-

guishes man from animal, so writing distinguishes civilized man from barbarian....
Writing exists only in a civilization, and a civilization cannot exist without writing. (1963,
221–22)

The relation between the invention of printing and
the origin of modern civilization is particularly important because the information explosion that resulted
from the invention of printing is the only one that can
be studied in any detail. Yet historians have curiously
neglected this critical invention. For example, W.H.
McNeill's monumental survey *The Rise of the West* contains only a single reference to Gutenberg, and that reference is in a footnote to a chapter on the Far East
(1963, 531). Eisenstein is one of the few scholars to
study the effects of the invention of printing, and the
almost complete absence of detailed study of those
effects surprised her. She writes:

What were some of the most important consequences of the shift from script to print?
Anticipating a strenuous effort to master a
large and mushrooming literature, I began to
investigate what had been written on this obviously important subject...[but] there was not
even a small literature available for consulta-
tion. Indeed I could not find a single book, or
even a sizeable article which attempted to survey the consequences of the fifteenth-century
communications shift. (1979, xi)

Eisenstein gives many reasons why historians tended
to overlook the importance of printing. She mentions,
for example, the difficulty of trying to assess the effects
of printing by looking for new titles (with presumably
revolutionary ideas) in the catalogs of books offered for

sale by printers in the fifteenth century. Such studies missed the point that the catalogs themselves were novel. It was not the titles that were revolutionary but the fact that books were being advertised for sale (1979, 168).

Eisenstein describes the importance of printing to events such as the Copernican revolution in astronomy and the comparable revolution in anatomy and medicine in the time of Harvey. She writes:

> We cannot…study all aspects of the past, and intellectual historians may be well advised to leave many inventions…to the specialist. To treat Gutenberg's invention in this way, however, is to miss the chance of understanding the main forces that have shaped the modern mind. (1979, 24–25)

The late Renaissance witnessed a number of singular and highly important events—the Copernican revolution, the discovery of the New World, the Reformation—that were crucial to the development of modern civilization. When these pivotal events are examined in detail a common thread can be found: books. As Gingerich put it:

> The early 1500s were times of vast changes. Oceanic navigators opened new continents and an age of exploration. Da Vinci and Dürer coupled mathematics with art to capture new harmonies of proportion and perspective. Martin Luther successfully set into motion a reformation of the church.… Meanwhile, the explosive spread of printing with movable type beginning in the 1450s fanned the sparks of all these movements, including the reform of astronomy. Without printing, Copernicus would have

been deprived of the vast majority of his source materials. Even five decades earlier, he could not easily have found the requisite information that built his *De Revolutionibus* into the greatest astronomical treatise of its century. And without printing, his manuscript might have languished, virtually forgotten, on the shelves of the cathedral library. (1975, 202)

To understand how widely Copernicus's book was distributed in the sixteenth century, Gingerich has made an effort to track down and catalog all the existing copies of it. He was able to find 245 copies of the first edition (1543) and a similar number from the second edition (1566), and he estimates that he has accounted for about half of the total printing. Thus about 1,000 copies were in circulation in the sixteenth century, and the copies had a wide geographic distribution. One of them had traveled as far as Mexico by 1600 (Gingerich 1979). With the invention of the printing press, books suddenly became available in numbers that had been beyond the reach of earlier civilizations; ideas were able to reach a more widespread audience than ever before. Copernicus's book is just one example of the flood of information that became available after the invention of the printing press.

The importance of printing to the Reformation is equally clear. Here the obvious critical book was the Bible, the first book to be printed and still the record holder for volumes printed. In addition, Martin Luther made extensive use of the new printing press. As Dickens put it:

Between 1517 and 1520, Luther's thirty publications probably sold well over 300,000 copies.... Unlike the Wycliffite and Waldensian heresies, Lutheranism was from the first the child of the printed book, and through this vehicle Luther was able to make exact, standardized and ineradicable impressions on the mind of Europe. (1966, 51)

For the discovery of the New World, the critical books were works on geography and natural history. Columbus, for example, owned at least four of the newly printed books. They are preserved to this day and contain extensive marginal notes in his own handwriting. A single generation earlier it would have been inconceivable that the son of a Genoese weaver could be wealthy enough to possess several books. But Pliny and Marco Polo were at Columbus's bedside. Morison (1942, 92–95) has outlined in some detail the importance of these books not only for the development of Columbus's ideas about geography, but also for his ability to argue effectively in the famous debates with the Spanish geographers at Salamanca, who in fact had a much better idea than Columbus did of the true distances to Cipango and Cathay.

Thus the explosive effect of the sudden and unprecedented production of an abundance of information (not all of it correct) provides a simple explanation for many of the unparalleled accomplishments of the Renaissance. It is an explanation that does not require assumptions of the presence of unusual human genius during this time period. The anomaly that was responsible for many of the accomplishments of the late Renaissance lay not so much in human intelligence as in the quantity of information that was suddenly available to that intelligence. The printing press supplied talented people with the information that they vitally needed and, equally important, allowed them to communicate their ideas and discoveries quickly to thousands of their contemporaries. Is it any wonder that the first generation to produce books in any quantity would explode in a frenzy of exploration? That the generation that first produced abundant copies of Pliny and Marco Polo would seek out strange climes and new worlds? That the generation that printed a thousand copies of *De Revolutionibus* would ask questions about the nature of the universe?

Eisenstein has raised questions about the quality of the information produced by the early printing industry, pointing out that much of what was printed was misinformation (1986, personal communication). Early books contained errors ranging from descriptions of unicorns and dragons to the inaccuracies in Ptolemaic astronomy. But the quantity of misinformation dropped rapidly following the invention of printing. Serious reports of dragons were rare by the eighteenth century. Why? At least part of the answer lies in the development of scientific methods in the sixteenth and seventeenth centuries. But this begs the question. Why were scientific methods suddenly developed?

Consider that scientific methods are fundamentally a two-pronged attack on problems of information. One part of the scientific method involves verification of information by observation and experiment (such techniques are obviously responsible for much of the decline in misinformation following the invention of printing). The other part of the scientific method involves a search for patterns or underlying rules ("laws") that explain large bodies of observations. This portion of the scientific method is, in essence, a technique for dealing with a large quantity of information by reducing it to a comprehensible pattern.

None of these techniques was particularly new. Observation, experimentation, and the discovery of patterns are probably as old as mankind itself. Yet the use of these techniques increased enormously in the sixteenth and seventeenth centuries, so much so that a "scientific revolution" is widely recognized to have occurred at that time. In other words, following the invention of printing we find, first, a massive increase in the production of information, and then an equally massive development and application of old techniques for refining that information. The scientific revolution of the sixteenth and seventeenth centuries can thus be seen as a necessary and even forced response to the large increase in the production of information that followed the invention of printing. The very quantity of newly produced information forced the development of techniques for dealing with large quantities of information. Those techniques are the processes that lie at the heart of modern scientific methods.

Eisenstein discusses a case that illustrates the direct impact of the sheer abundance of new information on the development of scientific methods in the sixteenth century. The case involves Tycho Brahe, who is rightly celebrated for producing superbly accurate new observations of the motions of the planets. His industrious pursuit of new data is frequently contrasted with "the 'almost hypnotic submission to authority' associated with reliance on Ptolemy and inherited data, manifested by all previous astronomers, including Copernicus" (Eisenstein 1979, 624). But the critical fact often overlooked is that Brahe possessed a printed set of Copernicus's newly published tables in addition to a newly printed copy of Ptolemy's tables. Brahe may have been the first person to *ever* possess two separate sets of computations based on two different theories (Eisenstein 1979, 624). And they did not match. Is it any wonder, then, that he ceased to rely on authority and turned to direct observation? What appears at first blush to represent a major and inexplicable change in basic philosophy can be seen as merely a natural reaction to the simple possession of an unprecedented quantity of information (none of it perfectly accurate) in the form of two sets of planetary tables.

These examples of the direct impact that the explosive growth in information production can have on the development of civilization do not, in themselves, provide a complete understanding of the quantitative aspects of the information limitations that marked the major watersheds of human history. To understand these information limitations we must first determine the quantities of information actually possessed by different levels of civilization. Then we must establish the minimal quantities of information that are required to perform certain tasks essential to those levels of civilization. Neither of these projects is easy. Both of them are possible only because the quantitative differences in the information requirements of different levels of civilization are so large that very rough estimates will prove to be both useful and revealing.

To construct these estimates we need to know a little about information theory. Like many pivotal discoveries, information theory is based on ideas so simple that they seem obvious rather than revolutionary. Claude Shannon was the first to recognize that the fundamental unit of information measurement is the quantity of information needed to decide between two alternatives. Shannon named this unit of information the "bit" (from BInary digiT). If one bit can decide between two alternatives, then two bits can decide among four alternatives by first dividing the four alternatives into two sets of two alternatives each. One of the bits then decides between the two sets, and the other one between the two alternatives within the selected set. Similarly, three bits can decide among eight alternatives, and so forth. Each additional bit doubles the number of choices that can be made. Because a letter of the alphabet is one of twenty-six alternatives (ignoring capitals), a letter contains a little less than five bits of information. This is only a small sample of information theory, but it is enough to allow us to make rough comparisons of the information content of various civilizations and thereby assess the effects of information explosions. It allows us to define an *information explosion* quantitatively as an increase of about two (or more) orders of magnitude in the production of information.

In order to evaluate the amount of information produced at various times in the past we need to distinguish five broad categories of civilization that differ principally by the method they use to store and handle information. It will be convenient to assign labels to them as follows:

Level 0—Pre-Language
Level 1—Language
Level 2—Writing
Level 3—Printing
Level 4—Computers

In evaluating the quantity of information available at each of these five levels it is the smallest quantity, the amount of information at Level 0, that is the most difficult to handle. Lacking language, each individual is essentially limited to the content of his or her own mind. But how much information can one mind contain? This quantity is so uncertain and variable that it might be best to simply label it h and try to set some bounds on it. We can set a lower bound on h by noting that epic poems such as the *Iliad*, which contain about 5 million bits, have been memorized. To get an upper bound, we might ask how many *Iliad*s a person might reasonably memorize. A hundred seems unlikely. It seems plausible that h is within one or two orders of magnitude of 5 million bits.

In a Level 1 society (with language) individuals have available the information content of their own minds plus that of the rest of their village, clan, or tribe, perhaps 50 to 1,000 times h There is, however, considerable redundancy in this information. The knowledge of how to hunt or chip flint would be common to many individuals. Even allowing for this redundancy the amount of information available to an individual in a Level 1 society would be one or two orders of magnitude greater than that available in Level 0.

In a Level 2 society (with writing) the amount of available information took another quantum leap. The greatest accumulation of information in any Level 2 society was probably found in the library of Alexandria in Egypt in about the third or fourth century A.D. One scholar describes Byzantine records indicating that the library possessed 532,800 scrolls in the third century B.C. (Parsons 1952, 204). Ignoring questions about the accuracy of this number, we could use it as the basis of a rough estimate of the information content of the library if we could determine the information content of a scroll. A clue to this value is found in the tradition that Zenodotus, a librarian at Alexandria, divided the *Iliad* and the *Odyssey* into twenty-four books each, in part so that each book would fit comfortably on a scroll (Parsons 1952, 205). Thus if the *Iliad* contains 5 million bits, and a twenty-fourth part of this is a typical size for a scroll, then the great library contained about 100 billion bits (10^{11} bits), one or two orders of magnitude more information than was available in a Level 1 society. Although the redundancy of the information contained in these volumes is difficult to quantify, it was probably less than the corresponding redundancy of the information available in a Level 1 society.

A Level 3 civilization (with printing) would have hundreds of libraries larger than the library at

Alexandria; some of its libraries are larger by several orders of magnitude. The total amount of information available even in an early Level 3 civilization is so vast that an individual cannot begin to comprehend all of it. Leibniz (d. 1716) is said to have been the last individual to comprehend all known information, certainly an exaggeration. In a fully mature Level 3 civilization the *daily* publication output can exceed any single individual's comprehension.

A crude estimate of the quantity of information available in a Level 3 civilization can be obtained by starting with an almanac figure showing 10,000 new book titles published in the United States in 1950. If each book contains 5 million bits, and 1,000 copies are printed, and the copies have a shelf life of about 20 years, then 10^{15} bits will be available in books alone. Allowing one or two more orders of magnitude for other publications such as newspapers, periodicals, government publications, advertisements, and so forth, we arrive at about 10^{17} bits of information available in a Level 3 culture, about a million times more information than was available in a Level 2 civilization (even to a scholar who had access to the library in Alexandria).

Level 4 civilization will be marked by a rapid acceleration in the rate at which information is produced. For example, word processors and computerized typesetting machines will greatly enhance the productivity of the publishing process, and computerized measurement and control devices will multiply the rate at which observations and other raw information can be generated. Modern accelerator experiments in high-energy physics, for example, are expected to generate a quantity of data equivalent to the content of the library of Alexandria in about five minutes, on average (Butler and Quarrie 1996). But Level 4 will see an interesting twist in the growth of information resources: Mere *increases* in the total store of information will become relatively unimportant, because even a Level 3 civilization can generate information at a rate that far exceeds anyone's ability to make use of it. Electronic computers, however, are capable of creating a totally new dimension in an information explosion. Computers can multiply our ability to find, analyze, and make use of vast quantities of extant information, thereby circumventing the information limits that bedeviled Level 3 civilization.

We can make some rough estimates of the amount by which computer technology can increase our ability to find and utilize information. We do not need very accurate estimates to show that the increase is going to be very large indeed. Consider the total quantity of information available to a talented person without a computer. Suppose this individual has completed a speed reading course and can read 1,000 words per minute. One word is about five letters or 25 bits, so if this person spends six hours a day reading, seven days a week, for seventy years, he will have read about 2×10^{11} bits (roughly twice through the library at Alexandria). A modest home computer can read that many bits in a few days—a really fast computer, in minutes. Therefore even today's computer technology provides an increase of a factor of thousands to millions in total information availability. This estimate makes no allowance for such things as the use of indexes to enable an individual to sort through more information than he or she can read, or for the fact that a human reader generally employs more intelligence and judgment than a machine is able to employ. But even with these refinements, it remains clear that the computer will increase information availability by many orders of magnitude. Increases on this scale are completely unprecedented and are far greater than the increase associated with the change from Level 2 to Level 3 civilizations.

In order to fully exploit the capability of computers to search through and analyze information in great quantities, information "utilities" are needed to supply information to home or office computers at modest prices. Rudimentary versions of such utilities were first created in the 1970s. The recent explosive growth of the Internet exhibits not only the technological capabilities that are possible today but also the incredible level of demand that already exists for such services.

And this is only the beginning. These services have been in existence only for a short while. Soon whole libraries and eventually virtually all of the information produced by our civilization will be available in this form. Computers will be able to search out needed information and distribute it electronically to the home or office. Every individual in a Level 4 civilization will have instant access to a supply of information that will dwarf even the Library of Congress, and will have the electronic hardware and software needed to make effective use of such a quantity of information. The ability to easily find and utilize the *entire* information stock of a civilization will be the hallmark of a Level 4 civilization.

Thus the quantitative limits on information available at each level of civilization are approximately:

Level 0—Pre-Language: 10^7 bits
Level 1—Language: 10^9 bits
Level 2—Writing: 10^{11} bits
Level 3—Printing: 10^{17} bits
Level 4—Computers: 10^{25} (?) bits

These estimates need refinement. They may be in error by orders of magnitude; the last one is little more than a guess. The relative sizes, however, are probably accurate enough to allow us to begin to assess the information limitations of the different levels of civilization.

Having defined these quantities, we are halfway toward an understanding of information limits. But the second part of the problem—identifying the tasks that are essential to a given level of civilization and estimating the amount of information required to perform those tasks—is considerably more difficult and will require an extensive study that can only be outlined here.

Consider the tasks that were actually accomplished at the various levels of civilization. Great literature, for example, has been created by civilizations at Level 1, in the form of epic poetry and sagas. Such literature can therefore be achieved by a civilization whose information limits are only a couple of orders of magnitude larger than the content of the literature itself. Operating a democratic system of government on a continental scale almost certainly requires the information capacity of the printing press, and is probably beyond the capability of a Level 2 civilization. And manned exploration of the moon is a task that exceeds the information capacity of a Level 3 civilization. The design, construction, and operation of a manned moon-rocket require the information processing capacity of electronic computers.

It will not be easy to construct a complete list of the features that characterize each level of civilization, the features whose information requirements we need to determine. A partial list for a Level 3 civilization would include such things as internal combustion engines, telephones and telegraphs, large-scale democratic governments, and universal literacy and education requirements. A similar list for Level 2 would include such things as large-scale autocratic governments, monumental architecture, and elementary mathematics. Level 1 would include such things as the use of tools, fire, and possibly agriculture and metallurgy. Obviously it will require a great deal of research and study to arrive at reasonably complete lists of features that characterize each level of civilization.

The quantity of information that is required to produce these particular features of civilization can be somewhat difficult to estimate, when both direct and indirect information requirements are considered. The information required to build an automobile, for example, goes far beyond the blueprints, diagrams, and specifications required to describe each part of the vehicle. The specifications for the size, shape, and strength of a piston are not very useful without information on how to obtain iron ore and produce steel. Despite these difficulties we should be able to estimate the direct information requirements of various tasks from the specifications and plans required for the task, and to make reasonable "overhead" assumptions that will account for the indirect requirements. Thus we should be able to make estimates that are sufficiently accurate to enable us to understand the limitations placed on civilization by limits on the amount of information produced.

To take a simplified example of such an estimate, let us try to calculate some of the information requirements for a large-scale democratic government system. Assume that we have a voting population of about 100 million, that two candidates are running for office, and that each has a 1,000-word (at 25 bits per word) statement of his or her views on critical issues. Further assume that each copy of the candidate's statement can be shared among ten voters. Simple arithmetic shows that more than 10^{11} bits are required for just two candidates in a single election (not even considering the indirect information requirements). This is roughly the size of the entire library of Alexandria, which represented centuries of accumulated information. It should be clear from this calculation that any large-scale democracy is going to have serious difficulties in a Level 2 civilization.

Similar calculations can be done for other critical elements of modern civilization. It should be easy to show that such things as communication networks, manufacturing complexes, and modern transportation systems have massive information requirements that exceed the information production capability of earlier civilizations. Imagine trying to run an airline without printed schedules and timetables, for example. Calculations of these information requirements are particularly important because they can be used to test, perhaps even falsify, the conjecture that civilizations are fundamentally limited by the quantity of information they can produce, store, and distribute. If, for example, it were shown that earlier civilizations had much larger information resources than I have estimated, or that the information requirements of modern civilization are much smaller, then it might be possible to demonstrate that civilizations are not really limited by restrictions on their information production.

Of course it is something of an oversimplification to say that civilization is limited by information, or, more generally, to say that civilization *is* information. But every statement that can be made about civilization is an oversimplification to some degree. The key point is not whether it oversimplifies, but whether this idea is sufficiently accurate and novel that it provides us with important new insights, as it does. Among other things, it provides a simple and natural explanation for vexing questions such as why the first industrial and scientific revolution occurred in western Europe rather than, for example, in the Arab civilizations. (In China, where the printing press was first invented, the concomitant information explosion was hampered by a number of factors, most obviously the absence of an alphabetic script.)

A number of other historical events are difficult to understand without a knowledge of quantitative information limitations. For example, the transition from ancient civilizations to the Dark Ages is usually described in terms of the destruction of classical civilization by barbarian invaders. But this explanation fails to provide a clear means of distinguishing between a civilized and a barbaric society, to explain why Theodoric the Goth, for example, should be considered more barbaric than Caligula or Commodus. From the standpoint of information limitations, however, the transition to the Dark Ages is seen as the most recent (possibly the last) example of the destruction of a Level 2 civilization by a Level 1 civilization. The few examples of the use of writing that are found during the aptly named Dark Ages (such as Bede, Charlemagne and Alcuin, and Alfred the Great) serve only to highlight the general absence of the written word that characterizes the information limitations of this period.

The medieval period is a time of gradual, halting progress back to a Level 2 civilization, complicated at the end by the first appearance of a Level 3 civilization. It is this double revolution that makes the Renaissance period difficult to understand and classify. The term "Renaissance" is itself a serious misnomer, for the Level 3 civilization that began here bears no resemblance to any classical civilization purportedly reborn.

One possible objection to classifying civilizations according to their quantitative information limits is that the levels defined are too broad—each level spans widely disparate types of civilizations. Is it reasonable, for example, to group Egypt of the twenty-third century B.C. with France of the twelfth century A.D.? Or are these cultures too dissimilar to be linked in any reasonable classification scheme? But an important function of a classification system is to point out connections that might not otherwise be obvious. The term "Bronze Age," for example, draws a link between Mycenaean Greece and Shang-dynasty China. And there *are* broad similarities between, for example, ancient Egypt and medieval France that make such a link reasonable. A visitor from Egypt's Middle Kingdom would have found much that was familiar in twelfth-century Paris: a hereditary monarchy; a warrior class that employed swords, bows and arrows and spears; and a religion that employed elaborate ceremony to ensure a pleasant life after death. The royal tombs at Fontevrault might surprise our visitor only in the use of bronze rather than gold for the coffin effigies. Such a visitor might even notice a parallel between the collapse of the "broken" pyramid at Meidum and the collapse of certain medieval cathedrals, notably the one at Beauvais. Engineering was strictly a trial-and-error affair in both cultures. The similarities between these Level 2 cultures appear far more striking than the differences, even though the two cultures are separated by thirty centuries.

If we were to continue this thought experiment by moving our Egyptian friend (or a medieval compatriot) into the next level of civilization we would find a radically different situation. Suppose we could move our visitor only eight centuries further, into a phone booth in an airport sometime around 1930. He or she would be in a truly alien world, filled with totally unfamiliar artifacts such as telephones, airplanes, automobiles, and refrigerators. Even a simple flashlight would have no parallel whatsoever in our visitor's experience—its function would seem like magic no matter how painstaking the effort to analyze it or work with it. (Arthur C. Clarke, the noted science and science fiction writer, has proposed as "Clarke's Third Law" that any sufficiently advanced technology is indistinguishable from magic. We can now quantify "sufficiently advanced," and note that it requires approximately one information explosion.)

One might argue that the twentieth century has some unique quality that makes this last comparison unfair or at least unrepresentative of a general rule. Yet an individual from a Level 1 civilization (from the central Amazon, for example, or from 10,000 B.C.) would have about as much trouble understanding the great Pyramid or the cathedral at Chartres as Aristotle or Roger Bacon would have trying to understand a flashlight. As a general rule civilizations within any of these levels resemble each other far more than any of them resemble any civilization in another level. This is the fundamental rationale for proposing such a classification system.

Of course the quantity of information available at each of these levels was never constant. In a Level 1 civilization the quantity fluctuated roughly with the size of the tribe or clan. In a Level 2 civilization the quantity of information produced was not limited in principle, but limits were determined in practice by the difference between the rate at which information was produced and the rate at which it was destroyed. The human penchant for burning books and libraries kept the limit fairly low throughout much of history—so low that Level 2 societies often had little advantage over Level 1 societies. Tales of the destruction of Level 2 civilizations by Level 1 are common in history.

Level 3 civilization was fundamentally different in this respect. After the invention of printing the production of information increased rapidly and never reached an equilibrium. In other words, the rate at which information was produced nearly always exceeded the rate at which it was destroyed. The technical advantage that

Level 3 civilizations held over earlier levels quickly became so great that I can find no recorded case of a Level 3 civilization being destroyed by one at Level 2 or lower. On the contrary, on each occasion in which such civilizations came in contact, as they did in the American continents and in India, the Level 3 civilizations invariably conquered or destroyed the earlier civilizations. Indeed, the first civilization to attain Level 3 very quickly conquered most of the world and relinquished control only after the rest of the world had also attained a good measure of Level 3 technology.

Note that I am not arguing for any form of technological determinism, that certain levels of information production would force the development of certain kinds of civilization. I am rather arguing the converse, that lack of information-production capacity will prevent the development of certain forms of civilization. Neither am I advocating a monocausal interpretation of history. Information production is only one element in the bewildering complexity of the development of civilization. Nevertheless, it is an important element, perhaps the most important element, and it is one whose time is ripe for study because the necessary theoretical tools have now been developed to a reasonable state of maturity.

What insights does an understanding of information limitations give us into the effects of the information explosion taking place today? Can we make any reasonable prognosis about the characteristics of a Level 4 civilization? Each of the three information explosions in the past produced a society that was largely unrecognizable to an individual from an earlier civilization. There is no reason to expect less from the fourth. In fact, this information explosion will be much larger and will take place much faster than the earlier ones. The difference between the twentieth and the twenty-first centuries may well be greater than the difference between the twentieth and the thirteenth (A.D. *or* B.C.).

Although making a forecast across such monumental change may seem hopeless, one historic figure did in fact make an extraordinary set of predictions that spanned just such different levels of civilization: Roger Bacon wrote about airplanes, submarines, horseless chariots, and other Level 3 wonders in the thirteenth century. Of course Bacon's remarkable prognoses were based more on wish than on hard evidence. Similarly accurate predictions about Level 4 civilization will probably be difficult to distinguish from wishful thinking.

One way to appreciate the full capability of a Level 4 civilization is to recognize that the Apollo moon landings were among the first (and presumably the simplest) examples of accomplishments that are possible *only* in a Level 4 civilization. What other tasks will lie within the capabilities of Level 4? Most of the problems facing our civilization today could be reduced or even eliminated with a massive influx of information. A partial list of those problems reads like a litany from the Apocalypse: famine, pestilence, poverty, war, illiteracy, intolerance. Level 4 civilization will be able to accomplish things such as observing cropland on a worldwide scale (from spacecraft) to monitor and alleviate the effects of drought, blight, locust plagues, and other age-old calamities. It will map the entire human genome (and probably many other genomes) and produce much better understanding of the functioning of biological systems. And these are just a couple of examples of the early accomplishments of Level 4. The full range is as far beyond our present imagination as our present civilization is beyond the medieval.

One of the most hopeful predictions that we can make about Level 4 civilization is that wars should be difficult or even impossible for Level 4 governments to conduct. War is already rare between two countries in which both have an unrestrained press. It is no accident that many of the wars fought in this century were launched by governments that kept tight control over the distribution of information, especially through the press. And such restraint on the press will be impossible in a Level 4 civilization, in which each home could have the equivalent of a printing press in the form of a small computer. War between nations in a Level 4 civilization may become as unthinkable as war between Virginia and Pennsylvania today. The antiwar movement could well match the achievement of the abolition movement of the last century.

Although Level 4 civilization will certainly solve some of the problems we face today, it will also undoubtedly have its own share of troubles. Crime, for example, will take on new and inventive forms, especially if electronic banking becomes standard. However, this discussion has focused on the gains and benefits conferred by computer technology rather than on the problems that will be experienced in a Level 4 civilization. It has done so for the most basic of reasons: Anticipating the problems that will occur in the future is a far more difficult task than forecasting the present-day problems that will be solved. The simple reason for the difficulty is that we generally have a deep familiarity with and understanding of today's problems that need to be solved, but at the same time we often lack even the context necessary to perceive the problems of the future. For example, although Roger Bacon wrote of airplanes, automobiles, and submarines, it would be preposterous to expect him to have anticipated, in the thirteenth century, the problems that would result from the successful attainment of his forecast, problems such as air pollution produced by automobile exhaust, acid rain, depletion of petroleum reserves, nuclear waste disposal, or deforestation of tropical rain forests. He lacked the necessary background and context even to conceive of these problems. As recently as the early twentieth century, Henry Ford did not anticipate the problems of automobile exhaust pollution, and Marie Curie would probably have been astonished at the problem of nuclear waste disposal.

There is therefore a broad variety of problems whose very context was technologically unimaginable only a generation or two ago. Consider the difficulties that would arise if parents had the ability to specify and alter the complete genetic makeup of their children. Such capabilities may become possible with computerized genetic engineering. This would not cause any great problem if, for example, all children turned out to have blue eyes, but what if all of them turned out to be male?

To take another example, computer technology raises serious problems concerning violations of personal privacy. Electronic mail can be intercepted, and electronic commercial transactions can be recorded and analyzed in ways that were inconceivable a generation ago. However, in addition to creating this problem, computer technology has provided a possible solution to it through the use of recently invented codes and ciphers that are believed to be mathematically unbreakable. A full discussion of the theory of these codes would take us far afield of the main point here. Good discussions can be found in Gardner (1989) and the references therein.

But perfect codes and ciphers, while they may alleviate problems related to privacy violations, can create other problems that may be more serious. Indeed, these codes may represent one of the most dangerous inventions of the century. Suppose, for example, they had been widely available during the Second World War, and

Allied code breakers had been unable to break the Axis codes. The battle at Midway may well have been lost, and the war in Europe could have dragged on for years, perhaps into another Thirty Years' War. Or more likely, the war would have dragged on until large numbers of atomic bombs were employed. Thus there are good reasons why many are afraid of allowing this technology to become widely available. It could prove extremely useful to criminal, terrorist, and drug-smuggling organizations, and correspondingly frustrating to law-enforcement efforts.

Level 4 civilizations will also have to face problems brought on by rampant population growth. These problems are intertwined with the computer revolution because computer technology is revolutionizing medical techniques and thereby decreasing death rates. Population growth problems are so important and so intractable that they warrant a full discussion in a separate chapter (see chapter 10).

The key point here is that all technological revolutions create problems that literally cannot be imagined outside of the context provided by those very revolutions. By failing to focus on these problems I do not mean to imply that they are unimportant, only that I lack the context necessary to even conceive of most of them. The one prediction that we can make with some confidence about Level 4 civilization is that its troubles will be different from the troubles that preoccupy us today. Problems such as the long conflict between capitalism and Marxism that troubled so much of this century may soon be as forgotten as the Schleswig-Holstein question that convulsed another century.

Today we can catch only dim glimpses of the features of Level 4 civilization, much as Roger Bacon caught glimpses of Level 3. And although there may not be anyone alive now with the insight and perspicacity to match the good friar, we have the advantage that our predictions need cover only a few years rather than centuries. Such is the staggering pace of the changes that we face.

2

"Theories of Everything" and the New Copernican Revolution

If the invention of the computer marks a genuine change in the level of civilization then we should expect it to be accompanied by scientific revolutions that produce deep and fundamental philosophical innovations, radical changes in our understanding of the universe and in our relation to it. Certainly the Renaissance period was marked by just such a fundamental transformation of our understanding of the universe. Freud once commented that the greatest of these scientific revolutions have, as their only common feature, the dethronement of human arrogance from one pedestal after another of previous conviction about our own cosmic importance.

The archetypal scientific revolution of the Renaissance was launched by Copernicus in the sixteenth century, and it led eventually to the discovery of the size or scale of the universe itself. Today the development of computer technology and theory has led us to the edge of an even larger and more important revolution concerning the discovery of the size of the "universe" of mathematics. The parallels to the earlier revolution are so remarkable that it is not at all farfetched to think of this as a new Copernican revolution.

The key breakthroughs of this revolution were made

by Gödel and Turing in the 1930s, and continued with the development of chaos theory in the latter part of this century. Although the new revolution began in mathematics and computer theory it has broad implications for other specialties, including fields as varied as physics and philosophy. As Traub and Wozniakowski (1994) put it: "We believe it is time to up the ante and to try to prove there are unanswerable scientific questions. In other words, we would like to establish a physical Gödel's theorem." Jackson (1995) responded in the negative: "Logical limitations within…formal fields (à la Gödel, Turing, Church, et al.) do not establish a limitation to scientific knowledge." Contrary to Jackson's conclusions, I will argue that there *are* mathematically demonstrable limits to scientific knowledge, that it is possible to prove the existence of unanswerable scientific questions, and further that the theorem that causes the most trouble for physical theory is Turing's rather than Gödel's. Although Gödel's theorem does raise fundamental questions about the behavior of the physical universe, we will see that these questions are themselves unanswerable. Turing's theorem, in contrast, is directly applicable and demonstrates the existence of sharp limits to physical knowledge.

To understand this new Copernican revolution and some of its implications it may be helpful to sketch some of the history of both revolutions.

The revolution touched off by Copernicus did more than just remove the Earth from the center of the universe. It also gave the first hint of the scale, the sheer magnitude of the universe. Prior to Copernicus the Earth was thought to be not merely the center of the universe, but also a fairly large fraction of that universe. Of course to the pre-Copernican cosmologist there was a Heaven up there somewhere and a Hell down somewhere else. But the overall scale of the universe was thought to be comparable to the scale of the Earth itself, within perhaps a few orders of magnitude. The annihilation of this idea was one of the greatest conceptual and intellectual challenges faced by human civilization. The preponderance of the universe was found to be fundamentally beyond our reach.

Following Copernicus, our understanding of the scale of the universe proceeded in a series of familiar stages. In the eighteenth century the scale of the solar system was determined with diurnal parallax observations of a transit of Venus (Ball 1908), using observations from Tahiti by James Cook (thus resolving what was then called "the final problem in astronomy"). The scale of distances to nearby stars was determined by Bessel in the nineteenth century using annual parallax observations (North 1994, 414–20). The scale of the Milky Way galaxy was determined by Shapley early in this century by studying variable stars in globular clusters (Motz and Weaver 1995). And finally the scale of galactic distances was first determined with Hubble's celebrated measurements of cepheid variables in distant galaxies (North 1994, 512).

Today we know that, in terms of light-travel time, the scale of the Earth is light-milliseconds; the scale of the solar system is light-hours; nearby stars are light-years distant; the Milky Way galaxy is 10^5 light-years across; nearby galaxies are 10^6 light-years away; finally, distant galaxies are 10^{10} light-years away. The ratio of the scale of the visible universe to the scale of the Earth is about 10^{20} in length, or 10^{60} in volume.

These numbers would have staggered Copernicus. They are literally astronomical and utterly beyond human comprehension except in the most abstract fashion. The knowledge that the Earth is an inconceivably small component of the universe is the enduring legacy of Copernicus and his distinguished successors.

The new Copernican revolution demonstrated that the "universe" of mathematics possesses a scale that is even more vast. The word "cosmic" utterly fails to do it justice. "Cosmic" connotes scales that are extremely large, but finite. The new Copernican revolution involves scales that are much larger than the merely finite.

What we might call the pre-Copernican mathematicians (that is, mathematicians prior to about the year 1928) believed that they understood the scale of mathematics itself. They further believed that it was a scale that would be comprehensible to individual mathematicians. The shattering of this notion and the discovery of the true scale of mathematics has created a revolution far greater than the original Copernican revolution, both in its magnitude and in its import for physics and philosophy.

From our perspective today it seems fantastic that, well into the twentieth century, mathematicians believed that they were on the verge of finishing mathematics (just as Captain Cook was engaged in finishing up astronomy a century and a half earlier). In 1928 David Hilbert actually proposed a serious program to attain this goal. Hilbert's program collapsed in a sequence of some of the most incredible discoveries in the entire history of science, and the story of the collapse forms the nucleus of a new revolution. Once again mankind found itself dethroned from a sense of cosmic importance, this time in its ability to comprehend mathematics itself.

The full history of this revolution could be traced back as far as the Pythagorean discovery of irrational numbers (see chapter 4), but a better starting point would be the formalism of geometry developed by Euclid in about 300 B.C. Following Euclid, mathematics was believed to define absolute truth, based on three fundamental and unassailable pillars. First, mathematics was based on (unprovable) axioms that were believed to be absolutely true. Second, it was based on formal logic as formulated by Aristotle, which was also believed to be absolutely true. Third, by using absolutely true axioms plus Aristotelian logic, mathematicians derived theorems that were also believed to represent absolute truth (Nagel and Newman 1968, 4–5).

It seemed an unshakable edifice. It was first shaken by the discovery of non-Euclidean geometries in the nineteenth century. The key idea in non-Euclidean geometry is that Euclid's controversial "parallel postulate" could be negated and the resulting axiomatic structure would produce a consistent geometry, albeit a different one from Euclid's. Mathematicians then realized that many, perhaps all, of the axioms that they believed to be absolutely true could be taken as either true or false. Such changes, instead of generating incorrect mathematics, merely generated new branches of mathematics (Nagel and Newman 1968, 9–11).

In the middle of the nineteenth century mathematics received a second shock. Beginning with the work of George Boole and the discovery of Boolean algebra, mathematicians realized that formal logic, the second "pillar" of the mathematical structure, was just another algebra, a branch of mathematics based on axioms and theorems developed from those axioms (Nagel and Newman 1968, 40). And the basic postulates of formal logic were seen to be no more immutable than the axioms of Euclidean geometry. Indeed, today an entire branch of logic called "fuzzy logic" is based on violation of the Aristotelian rule or axiom of the excluded middle, the idea that every statement is either true or false. (Fuzzy logic should probably be called non-Aristotelian logic, by analogy with non-Euclidean geometry.) The rule of the excluded middle defines a two-valued logic system (true-false). But other branches of logic, using three or more values or even a continuum of values between true and false, can be defined consistently and comprise simply other branches of the algebra of logic.

Thus two of the three "pillars" of mathematics, the absolute truth of axioms and the absolute truth of formal logic, stood open to some question at the start of the twentieth century. This situation gave rise to Bertrand Russell's famous quote that "Pure mathematics is the subject in which we do not know what we are talking about, or whether what we are saying is true" (Nagel and Newman 1968, 13). To try to bring order out of this chaos, Hilbert proposed that three proofs were needed to set mathematics back on a firm theoretical foundation (Hodges 1983, 91). In a famous address in 1928 he asked for the following proofs:

1. That mathematics is complete, so that every proposition can be proved either true or false.

2. That mathematics is consistent, so that if it is possible to prove that proposition A is true, it is not simultaneously possible to prove that it is false.

3. That mathematics is decidable, in other words that there exists a definite method that can be applied to every mathematical proposition and that is guaranteed to prove the proposition either true or false.

Hilbert saw this program as a matter of "tidying up" the foundations of mathematics. He thought that the proofs would be difficult and might take a long time to work out, but he had little doubt that all three propositions would be proved true. But to everyone's astonishment all three problems were resolved within a decade. To their even greater astonishment, all three propositions were essentially disproved. Kline observed: "Even Hilbert may not have foreseen the maelstrom that was to follow.... As Frege put it, 'just as the building [of mathematics] was completed, the foundation collapsed'" (1980, 196–97).

Kurt Gödel's celebrated incompleteness theorem settled the first of Hilbert's questions and established that mathematics is essentially incomplete, so that there are always infinitely many propositions that are both true and unprovable. Gödel further showed that, although mathematics must be consistent, this consistency could never be proved and must rather be accepted as an axiom. Finally, Alan Turing's theorem on the halting problem disposed of the third question by showing that mathematics was not decidable in Hilbert's sense.

The first of Gödel's demonstrations is critical for understanding the scale of the edifice of mathematics itself. Axioms are the fundamental components of mathematics. And the number of axioms required for all of presently known mathematics is not large. Geometry is unusual in requiring as many as twenty of them. Arithmetic needs only five, the Peano axioms (Nagel and Newman 1968, 103). The number of axioms needed for all of presently known mathematics is probably in the range of fifty to a hundred. But by Gödel's theorem, we now know that the number of axioms needed for all of mathematics is not a finite number. Further, Gödel's theorem shows that even an infinite number of axioms is not sufficient to contain all of mathematics. Gödel's theorem is constructive: Given any collection of axioms, finite or infinite (beyond some minimum number), it is *always* possible to construct propositions that are both true and unprovable, which can then be taken as new axioms (Hofstadter 1979, 465–71).

This is the essential core of the new Copernican revolution, that even infinite amounts of information (in the form of axioms) are not sufficient to span all of mathematics. Yet it is possible for us to discover and know only a finite number of axioms. Thus everything that it is possible for us to know of mathematics is (literally and precisely) an infinitesimal fraction of mathematics itself. In contrast to the conceptual change engendered by the first Copernican revolution, the scale of the Earth relative to the entire visible universe is enormous in comparison to the ratio of knowable mathematics to all of mathematics.

Gödel's work has fundamental implications for physics and philosophy as well as mathematics. We know now that there can be no "theory of everything" (TOE) in mathematics. But is there one in physics? In other words, is there a model consisting of a finite set of equations that explains all of the properties of matter and energy and their interactions? Only two possibilities exist, either there is such a theory or there is not.

Although these two possibilities seem distinct and radically different, in fact there is no practical or scientific difference between them. To determine the difference between them, or in other words to verify that a particular mathematical model is in fact a TOE, one would have to compare it with experimental evidence with infinite precision. This is an impossible requirement, yet nothing less will do. If the precision of the experimental verification were less than infinite, then the possibility would always exist that the proposed TOE was merely a good approximation to the behavior of the physical universe. And the situation is even worse than this. Carrying out an experiment to an infinite level of precision is a necessary condition for verifying a TOE, but it is not a sufficient condition. The sufficient condition would be to demonstrate that the TOE matches *every possible* experiment to an infinite precision. This requirement is completely absurd, far beyond physics and deep into the realm of pure fantasy.

Therefore there can be no practical difference for physics between a universe that contains a TOE and one that does not. In either case, candidate theories would have to be tested forever, or until they fail. (This situation bears a striking resemblance to Turing's halting problem, and, as we shall see, Turing's theorem places tight restrictions on the capabilities of physical theory.) Which of these two possibilities is correct is therefore a question that is beyond science. It is more a matter of theology, for it cannot be answered with scientific techniques. The only difference between a universe in which a TOE exists and one in which it does not is that in the first universe physicists will eventually become frustrated at their inability to find "new physics." They would probably then conclude that the "theory of everything" has been found, as indeed they have on at least two occasions in the past. But in neither universe is there any way to know for certain.

It might be argued that there is another way to determine the existence of a TOE, other than by comparison with experiment. It may seem absurd, but if a mathematical model could be shown to be the only one possible for physics, it would have to be accepted without recourse to experimental verification. There was at least one occasion in the past in which this was thought to have happened: It was once believed that Euclidean geometry was the only mathematically possible model for the structure of ordinary space. This idea was shattered by the discovery of the non-Euclidean geometries and their application in the general theory of relativity. In light of the essential incompleteness of our knowledge of mathematics, it is inconceivable that such certainty will ever be approached again.

But even if a TOE does exist in physics, we can show

that it will be less useful than we might think; there must still remain questions that even a TOE cannot answer. There is another remarkable parallel to the Copernican revolution here. The discovery of the scale of the universe took a strange twist in the early twentieth century with Einstein's development of the special theory of relativity. Not only is the universe incredibly larger than Copernicus had imagined, but Einstein showed that there is a completely unexpected barrier to exploring this universe: a universal speed limit. No more unexpected and counterintuitive discovery has ever hit physics.

But chaos theory (which might be broadly defined as the general study of nonlinear problems) provides a stunning parallel: A similar sharp limit constrains our ability to explore the infinite universe of mathematics. This novel restriction is not a limit on velocity, which indeed would have little meaning in the exploration of mathematics. The restriction is more direct and surprising: There are very few problems that mathematics can solve at all. The class of problems that can be solved is essentially limited to linear problems, plus a small number of very special nonlinear problems. The special characteristic that generally distinguishes the solvable nonlinear problems is a high degree of symmetry. The general study of nonlinear problems has shown that a fundamental difference exists between the solutions to linear and nonlinear problems: The solutions to linear problems can generally be expressed with a finite amount of information, and the solutions to nonlinear problems generally cannot, except in cases that involve very special symmetries. Chaos theory, in a broad sense, concerns the study of these very simple problems whose solutions are infinitely complex.

There are two general properties of the solutions to nonlinear problems that create the requirement of infinite quantities of information. The first is that the solutions generally do not repeat or form simple patterns (in this property there is a striking parallel to the non-repeating behavior of digits in the decimal expansion of irrational numbers). Thus the complete solutions to nonlinear problems must be specified in some fashion over infinite ranges. The second general property of nonlinear problems is that they are frequently extremely sensitive to changes in initial conditions. In other words, small errors in the solutions tend to grow exponentially. The only way to avoid this problem is to specify the solution with infinite precision.

Thus the difficulty is not just that nonlinear problems are more difficult to solve. Nor does it reflect our lack of cleverness in finding solutions. The problem does not lie in limitations to our technical and intellectual capabilities but rather in the characteristic properties of the solutions themselves.

We can illustrate the difficulty in Newtonian mechanics. The two-body problem is a classic example of a nonlinear problem with powerful symmetries that allow the solution to be attained. The particular symmetry involved in this case is periodicity in time. In the two-body problem the solution is an ellipse that can be specified for all time by just six parameters, because of the periodic nature of the solution. And if there is an error of, say, two centimeters at one point in the modeled ellipse, the error will remain two centimeters at that point forever. In contrast, the solutions to the general three-body problem never repeat anywhere. The general solution to the three-body problem must be expressed in the form of an ephemeris that tabulates the positions of the three bodies forever. And, as noted above, small errors in the ephemeris, instead of remaining constant, will generally grow exponentially with time. Thus no ephemeris can be useful except over a fairly narrow range of time, unless the ephemeris is infinitely accurate at all points.

What is extraordinary about this discovery is not just that essentially all of mathematics (and by implication, essentially all of physics) is out of reach, in theory. We already knew that from Gödel's theorem. The shock is that the insuperable difficulties arise in incredibly simple problems. Mathematics gets into serious trouble not with numbers that approach infinity, nor even with cosmic-scale numbers like 10^{20}, but with numbers in the range of 2, 3, and 5. Second-order nonlinear differential equations (e.g., turbulent fluid flow), three-body problems, and fifth-order polynomials all exhibit the level of complexity at which the problem is commonly found. In a broad sense (and with numerous exceptions) the realm of the solvable in mathematics is restricted almost exclusively to the set of linear problems plus excruciatingly simple nonlinear problems.

The preceding discussion is necessarily a bit vague—I have not carefully defined what is meant by a linear problem. The definition will differ depending on which branch of mathematics is being discussed. A complete description would be too extensive to attempt in this chapter and would have many qualifiers and exceptions. Nevertheless it is broadly correct in fields as diverse as, for example, algebraic equations, differential equations, and optimization theory, that linear problems are solvable, and nonlinear problems are intractable. The difficulty levels in mathematical problems span the entire range from problems that have exact, analytic solutions, to problems that can be approached only by numerical approximation, to problems such as fluid flow at high Reynolds numbers that are not tractable to numerical approximation even with the number-crunching power of modern computers.

We should not think that a sharp divide exists

between solvable and unsolvable problems in mathematics. The divide is real, but from what we know of it, its shape appears to be complex and fractal-like. For example, there are numerous examples of solved problems deep within the range of Newtonian N-body problems. The orbits of N bodies orbiting at the vertices of a regular polygon in a plane can be circular and repeat exactly, even for values of N approaching infinity. Another example is found in Lagrange's elegant solutions to the restricted three-body problem. And although there is no general analytic solution for polynomials of order greater than four, all twenty-four solutions to the equation $x^{24} = 16,777,216$ can be written down by inspection ($2.0\ e^{(i\,n\,\pi/12)}$, $n = 1, 24$). The symmetries that allow solutions to be attained in nonlinear problems are scattered with both regularities and irregularities that are reminiscent of the shape of the boundary of the Mandelbrot set. The Mandelbrot set could serve conceptually or schematically as a model (albeit in two dimensions) for the shape of the boundary of the domain of solvable problems. And the boundary is further fuzzed up by the use of methods such as perturbation techniques, which allow us to extend attainable solutions a short distance into the range of unsolvable problems.

The irregularities of the boundary between solved and unsolved problems in mathematics may have seemed to reflect nothing more than the accidents by which individual problems have been studied. But now these irregularities seem to represent something deeper, a real structure in the complex fabric of mathematics. And the hope that *all* of the intervening pieces of the fabric, the spaces between the solved problems, will eventually be filled in as more problems are solved, is fundamentally vain. Nearly all problems in mathematics must remain forever unsolved, even to a group of gifted and hardworking mathematicians working for an eternity. Eternity is inadequate to deal with the problem. This is the enduring legacy of Gödel's incompleteness theorem.

Therefore even if a TOE for physics exists and is known, if it contains any nonlinear elements, then mathematics is probably not a powerful enough tool to allow us to handle it. But another critical theorem places even more fundamental limits on our ability to understand the physical universe, limits that do not depend on assumptions of nonlinearity. The theorem is not Gödel's, but Turing's theorem on the halting problem.

In the 1930s Alan Turing studied the properties of general-purpose computing machines. He showed that a very simple computer, now called a Turing machine, can perform any operation that can be defined in mathematics. (Modern electronic computers are highly sophisticated implementations of Turing machines; any calculation that can be performed on a modern computer can also be performed on a Turing machine.) Turing then asked a very fundamental question: Given a computer program running on a Turing machine, is there any way to tell whether the program will halt or will run forever? To take some trivial examples, a program designed to multiply 7×2 will halt with the answer 14. But a program that calculates successive decimal digits of π until it finds an end will never halt. It is therefore possible in specific cases to determine whether a program will halt or not. But Turing wanted to know if a general algorithm exists that would answer this question in a finite number of steps for any finite-length program. Turing realized that the "halting problem" is equivalent to Hilbert's "decidability problem" (Stewart 1992, 301–5).

Turing answered the question by proving that no "halting algorithm" exists. The proof is closely related to

Gödel's (for a discussion of the relation between the proofs, see Casti 1992, chapter 9 and Stewart 1992, chapter 22). A halting algorithm, if it did exist, would be a mathematical tool of enormous power. Among other things it would immediately resolve almost every question in number theory. Fermat's last theorem or Goldbach's conjecture, for example, could be proved or disproved by writing a trivial program that would test all possible combinations of integers and halt when it found a counterexample. By itself, such a program does not help much, although it might settle the problem by finding a counterexample. However, if the theorem or conjecture is correct and no counterexample exists, then we would learn nothing in any finite time, because the program would run forever without telling us anything.

This is where the halting algorithm comes in. If such an algorithm existed, it could operate on this program and tell us in a finite number of steps whether the program would halt and therefore tell us whether or not a counterexample exists. We would never actually have to run the original program; we would only run the halting algorithm. This is the sense in which the resolution that Hilbert expected for his "decidability problem" (that such an algorithm does exist) would have finished mathematics itself (Stewart 1992, 303–6; see also Hofstadter 1979, 425–29). As G.H. Hardy put it:

> There is of course no such theorem, and this is
> very fortunate, since if there were we should
> have a mechanical set of rules for the solution
> of all mathematical problems, and our activi-
> ties as mathematicians would come to an end.
> (Hodges 1983, 93)

The nonexistence of a halting algorithm places fundamental limits on the ability of mathematics to make use of a TOE in physics, even if such a thing exists. (Ian

Stewart pointed this out in a recent issue of *Scientific American* [1994].) Consider, for example, the famous cellular automaton game called "Life," invented by John Conway (see Gardner 1983 for a discussion of the game). This game provides a model "universe" in which the TOE is perfectly known; indeed it consists of only three simple rules.

"Life" is played on a square lattice (a checkerboard) of infinite extent. Each square on the board can be either filled or empty. The game is started by filling an arbitrary set of squares. The set of squares that are filled on the next step is determined by the following three rules:

1. Any square that is empty will be filled if exactly three of its neighboring squares are filled (the "neighbors" are the eight squares that touch any given square).

2. Any square that is filled will remain filled if exactly two or three of its neighboring squares are filled.

3. All other squares are empty on the next step.

As the game progresses a surprising array of patterns frequently develops, often from very simple starting configurations. Conway noted that four possibilities exist for any given configuration of the game "Life": The configuration could die out; it could stabilize; it could oscillate; or it could grow without limit. He then asked if any algorithm would tell whether or not a given configuration would grow without limit. The question was answered by showing that a universal computer (a Turing machine) could be constructed on the board of the game "Life." The question of whether this configuration would grow without limit thus became equivalent to Turing's halting problem, and Turing's theorem tells us that no algorithm can answer the question.

The significance of this, as Stewart noted, is that even in a universe in which a TOE exists and is perfectly known, there still remain simple and intrinsically interesting questions that cannot be answered. The only requirement is that the TOE be sufficiently complicated to allow the construction of a Turing machine. Thus the key limit on our knowledge of the physical universe is set more by Turing's theorem than by Gödel's. The very existence of a Turing machine establishes the existence of unanswerable scientific questions, and thus resolves the question posed by Traub and Wozniakowski (1994).

We have reached an interesting dichotomy here: Having just discovered that the "universe" of mathematics (and probably physics) is too vast for us to ever be able to explore any significant fraction of it, we have at the same time invented the most powerful instrument ever devised for exploring it, the computer. Our position is perhaps somewhat analogous to that of a person who has just invented a spaceship capable of voyaging to nearby stars, and at the same time has discovered just how vast the universe really is. Distances to nearby stars are still insignificant compared to the scale of the universe. Like this intrepid explorer, we know now that although the preponderance of the universe of mathematics remains forever out of reach, still there remain many new worlds for us to explore.

The computer will enable mathematicians and scientists to explore the universe of mathematics to limits that are many orders of magnitude beyond the corresponding limits that could have been explored without this invention. Consider, as a very simple example, the calculation of the successive decimal digits of π. In 1873 William Shanks calculated the first 703 digits of π by hand. Perhaps even more amazing, he got the first 526 of them right (Dunham 1990, 111). With modern computer technology the number of computed digits of π is somewhere in the vicinity of billions. Thus the investigation of the digits of π has already gone six orders of magnitude further than was ever done by hand. And there can be little doubt that it will be carried many more orders of magnitude further in the future as computer technology advances.

Yet in a very real sense the calculation of π to a billion digits is no closer to being "finished" than is the calculation to 526 digits. The number of digits that remain to be calculated has not changed in the slightest. On the other hand, there are some fascinating questions about the statistics of the digits of π whose answers can be better approximated with a billion digits than with 526. For example, does each digit or combination of digits occur with the same frequency? And are the digits in π more random or less random that the corresponding digits in the square root of 2? Interestingly, it appears that they are more random (see the discussion in Seife 1997). There are innumerable questions like this that can be tackled with computer technology whose answers would remain forever unknown without that technology.

Thus, even though we now have extraordinary and unprecedented power for exploring mathematics, we also know now that no TOE can exist in mathematics, that there is not even an infinite collection of axioms that is adequate to contain all of mathematics. And our power to deal with this inconceivable complexity remains largely limited to the realm of linear problems (with a variety of important exceptions). The physical universe may or may not mirror the infinite complexity of mathematics, but even if it does not, and the laws governing the physical universe are simple and finite in number, there is absolutely no way of determining this fact with certainty. Finally, even if the underlying laws governing the physical universe are finite in number, and they are completely known, and they are as simple as, for example, the rules for the cellular automaton game "Life," then there must *still* exist simple, natural, and interesting questions that cannot be answered by any theory.

This is the new Copernican revolution, and it is as far removed and as strange to the concepts of mathematicians and physicists of the early twentieth century as the modern astronomical universe would be to Copernicus.

My personal preference is to believe that physics has this fractal-like complexity in the structure of its fundamental laws. In other words, I tend to believe that no finite or even infinite set of laws or equations are sufficient to describe the physical universe, and that progress in physics will consist of an unterminating sequence of successively better approximations to this infinite complexity. There are three rather unscientific reasons for this belief, the first being the somewhat amusing history of "theories of everything" in the past, from the Pythagoreans to the Newtonians. More important, as Eugene Wigner famously noted, mathematics is "unreasonably effective" in its application to problems in physics. If we assume that this unreasonable effectiveness will continue through the indefinite future, then it is hard to see how research in physics would end while progress in mathematics is still ongoing. And third, the universe appears to have an affinity for fractal-like complexity even in phenomena that are simple enough for me to perceive and understand. Clouds, trees, and coastlines have fractal-like properties. Why should the universe not reflect this complexity in its fundamental underlying structure? I recognize that there is a diversity of possible answers to this question.

Some will find the idea of an infinitely complicated universe frightening, even as some people very reasonably find the vastness of astronomical space frightening. I personally find it to be full of hope rather than fright to think that there will always be something new to explore and discover no matter how long the human race lasts or how far science progresses. Although the search for a TOE in physics is an exercise in perpetual futility (with mathematical certainty), the search for the theory that will provide the next half-dozen orders of magnitude improvement over our current understanding is one of the most exciting things that ever happens in science. It is comforting to think that future generations will never be deprived of this excitement.

Thus the fact that we can never attain complete knowledge of mathematics (or physics) does not mean that we have no motive for further progress. The situation is exactly the same as the classic aphorism about travel, that the journey is frequently more interesting than the destination. The journey through mathematics is going to be so fascinating that we should not be greatly disturbed that the final destination lies forever out of reach. The following two chapters describe something of the journey that lies ahead, as well as some of the many fascinating discoveries that have already accompanied the computer revolution.

3

The Computer Revolution
in Science and Mathematics

In previous chapters we saw parallels to the Copernican revolution in astronomy and to the revolution caused by the printing press. Yet another parallel revolution is going on today, a revolution in science and mathematics brought about by the invention of the computer. This one is a close analogue to the "scientific revolution" of the seventeenth century and also has strong parallels to the first flowering of mathematics in Hellenistic times.

The scientific revolution in the Renaissance began in astronomy with the work of Brahe and Kepler. It essentially invented physics, with Newton's development of calculus and classical dynamics, and it then gradually spread into other areas of science. In sharp contrast, the new revolution is going to happen simultaneously in almost every area of science and mathematics. This is completely unprecedented. Never have so many revolutions occurred all at once. Yet there are important parallels to the separate parts of the current revolution.

In the history of mathematics and science, three developments were so important that they did far more than just create progress in the field—they essentially defined the field for centuries following. The first was Euclid's formulation of geometry in the second century

B.C.; the second was Newton's invention of calculus and dynamics in the seventeenth century; and the third was the invention of the computer in this century. Of these three, the third is by far the most important, the most powerful, the most all-encompassing. To understand why this is so we need to examine the parallels to the earlier revolutions.

Both of the earlier revolutions had a similar structure, based on two key elements. The first element was a critical set of information or data. The second and more important element in both revolutions was an extraordinary advance in the analysis of those data.

For Euclid the critical set of information was the large body of knowledge about basic geometry that had been compiled irregularly over the centuries, largely in Egypt and Mesopotamia. The historical records of mathematics prior to Euclid are sadly incomplete, but it is clear that a great deal was known about geometry before Euclid's time. For example, the Pythagorean theorem had been known for some time, Thales had shown that the angles of a triangle summed to 180º, and Hippocrates had even managed to "square" the lune, a concave figure bounded by circular arcs (Dunham 1990, 1–26). Euclid's genius lay in a novel analysis of those known facts. He was able to show that through the use of Aristotelian formal logic all of known geometry could be derived from twenty simple postulates.

In the revolution of the seventeenth century the critical data were the astronomical observations that had been made in Denmark by Tycho Brahe in the late 1500s. Brahe's data were the first systematic astronomical observations taken in Western civilization since Babylonian times, and were the most important observations made in astronomy prior to the invention of the telescope. Lacking a telescope, Brahe made measurements of planetary positions that were accurate to about one minute of arc, the limiting resolution of the naked eye. As luck would have it, one minute of arc was good enough, barely.

Brahe's data were first analyzed by Johann Kepler using conventional geometric methods that would have been reasonably familiar to Euclid. And Kepler made a number of remarkable discoveries; perhaps foremost among them was that the orbit of Mars does not have the shape of a circle. The misfit to a circle was small, the famous "eight minutes of arc," but it was larger than the uncertainty in Brahe's data. Kepler then discovered that although a circle could not fit the data, an ellipse fit it rather nicely. Kepler's analysis pushed Euclidean techniques to their limit.

The revolutionary analysis of Kepler's results was made, of course, by Isaac Newton. Newton did far more than just resolve the problems posed by Kepler's results. His solution, in the form of the calculus and Newtonian mechanics, was so novel and so powerful that it seemed for a long time that it might solve *all* of the problems in science and engineering. Indeed, for several centuries following Newton, scientists and mathematicians believed that they were in possession of the final "theory of everything."

Today the elements that created the earlier scientific revolution are once again in place, but this time on a far grander scale, and all of them are critically dependent on the presence of computer technology.

Analogues to Brahe's data exist today in the form of vast digital databases. There is essentially no field of science that is not in possession of enormous quantities of data in this form. Most of these data sets were not even conceivable in the precomputer era. Examples include such things as DNA sequence data in biology, particle accelerator data in physics, digitized astronomical databases, and global seismic array data in geophysics. Even in fields as far removed from science as English literature, computer technology has allowed novel insights from such things as computerized analysis of the frequency of occurrence of word patterns in Shakespeare's work.

Thus vast quantities of new and vital data are extant today, just as in the 1600s. But now as then, data alone do not produce a revolution. It is new and powerful analytic techniques, new forms of mathematics, *new ways of understanding what the data are telling us*, that can and will produce such a revolution. And computer technology has allowed the development of a broad variety of remarkable and powerful new analytic techniques. These new techniques are important for exactly the same reason that the development of calculus was important: Both of them make previously difficult problems easy, and previously unsolvable problems solvable. And the principal difference with the seventeenth century is that the class of problems that are made easy or solvable by computers is vastly larger than the class that was solved by calculus.

Calculus is most useful in areas that are referred to as analysis, which is broadly the study of curves or functions that are well behaved, continuous, and smooth. And computer technology has allowed the development of symbolic algebra algorithms that will revolutionize the study and application of even these conventional mathematical problems. But computer techniques facilitate the study of curves that are not smooth and may even be discontinuous (fractals). More importantly, computer techniques also apply directly to many fields that calculus touches only indirectly, such as finite mathematics and combinatorics. Turing posited that every definable operation in mathematics can be performed by a computer. Turing's thesis makes plain the extraordinary breadth of the applications of computers in mathematics.

The new analytic techniques that are being developed with computers fall into two broad classes. The first consists of techniques or procedures that have been known for some time, but whose application was difficult, limited, or impractical without the use of computers. The other class consists of techniques that are original and new. As might be expected at this early stage of the computer revolution, the former class predominates today. But the second class is of more fundamental importance for the long term.

Examples of the first class of techniques include the fast Fourier transform (which was known to Gauss), numerical integration, Monte Carlo methods, and statistical techniques such as least-squares fitting. Today a least-squares analysis that solves for several thousand unknown parameters can be performed in a few minutes. This is a task that would have exceeded the capabilities of even Euler or Gauss.

Laplace once commented that the invention of logarithms had doubled the life span of an astronomer (Dunham 1990, 157). He meant, of course, that logarithms made calculations so much simpler and faster that an individual could finish far more work in a single lifetime. Consider how much more strongly Laplace's comment applies to the computer revolution, and how much greater is the difference. Supercomputer calculation speeds today are approaching one teraflop, or 1,000 billion-floating-point operations per second. A floating-point operation typically involves multiplying a pair of sixteen-digit numbers. With one teraflop capability you could complete in a single second a calculation that would take about 700,000 lifetimes of hard work, assuming that you are able to do a sixteen-digit multiplication in about five seconds, without error. This is more than just a quantitative change in computational capability; a quantitative change of this magnitude implies a large qualitative change in the class of problems that can be addressed and solved.

One important area in which the application of computer techniques has produced stunning results is the study of fractals, curves that may be continuous but are not smooth, such as the famous "Mandelbrot set" (see Dewdney 1985, or Briggs 1992). A number of mathematicians in the precomputer era, including Poincare, Julia, Koch, and Cantor, approached the fringes of fractal theory with the discovery of strange attractors, Julia sets, Koch "snowflakes," and Cantor "dust." And many of their contemporary mathematicians reacted in horror to the bizarre behavior of these novel constructions, in part because they could perceive the enormous computational effort that would be required for further progress in the area. What these somewhat myopic mathematicians failed to foresee is that these strange new constructions would have broad applications in the physical and biological sciences. Phenomena as diverse as clouds, coastlines, blood vessels, and lungs have been found to possess properties that are usefully modeled as fractal structures.

As noted above, examples of the second class of algorithms (i.e., techniques and methods that are completely new and unique to the computer age) are a bit less common at this early stage of the computer revolution. But they illustrate the type of progress that can be expected. Perhaps the most important new analytic technique so far developed is an algorithm that can evaluate indefinite integrals, one of the most notoriously difficult problems of the calculus. Pavelle et al. comment:

Such problems [indefinite integrals] are encountered often in the physical and biological sciences, but their difficulty has bedeviled mathematicians for hundreds of years. It was once thought that no general algorithm for the solution of such problems could be constructed.

But such an algorithm has been discovered and implemented. It involves a novel method for determining the general form for the solution. The full solution is then worked out backward by differentiation. The method is too computationally intensive to be practical without computers. In an amusing sidelight, the new algorithm was used to check a set of eight widely used tables of integrals. This check did not merely find errors: It found a lot of them. The overall error rate was around 10 percent, and in one of the tables it approached 25 percent of the published results (Pavelle et al. 1981, 145).

Other new analysis techniques that could not be implemented in the absence of computer technology include new optimization algorithms such as simulated annealing algorithms (Press et al. 1992, 436–49) and genetic algorithms (Rawlins 1991).

Some computer analysis techniques provide completely new ways of looking at problems. One of the most familiar is the cellular automaton. In a cellular automaton an n-dimensional space is divided into cells, each of which may contain a variety of defined objects. There is also a set of rules that define the interactions among the filled cells. The best-known example of a two-dimensional cellular automaton is the game "Life," invented by John Conway and popularized by Martin Gardner (1983) in a series of articles in *Scientific American.* Cellular automata have provided insight into processes as varied as the generation of color patterns on seashells and the limitations on "theories of everything" (as was noted in chapter 2).

Perhaps the most important effect of computers on science will lie in an entirely new avenue of scientific investigation that is being developed. Traditional science is divided roughly between theory and experiment. Today, computers have created a third branch of science—simulation, which is considered by many to be on a par with experimental and theoretical methods. The ability to undertake the massive calculations needed to simulate the behavior of the physical world has already led to important insights into processes as varied as the explosions of supernovae and the mechanisms of earthquakes. Simulation is an important extension of both theory and experiment, and its development is one of the pivotal events in the history of science. The development of a new branch of science is something that not even the invention of calculus can claim (although, of course, the calculus enormously enhanced the theoretical side of science).

In addition to a third branch of science, computer technology also has created a second branch of mathematics—experimental mathematics (Stewart 1992, 313–17). In one sense experimental mathematics is not new. Great mathematicians have frequently done experimental calculations to develop insights in preparation for the development of more general results and proofs. But the scale of the experiments has changed. The numerical experiments that are used to explore fractals, for example, exceed the capabilities of the greatest mathematicians. Other early results from experimental mathematics include the refutation of a conjecture by Euler, who suggested that at least n nth powers would be required to sum to another nth power (three cubes, four fourth powers, etc.). But Frye was able to find a

counterexample by using a combination of conventional analysis and a computer search (Stewart 1996, 36):

$$95,800^4 + 217,519^4 + 414,560^4 = 422,481^4$$
$$(= 31,858,749,840,007,945,920,321)$$

Other counterexamples are known, but this one is the
simplest possible case for fourth powers. Finding these
counterexamples exceeds the capabilities of even a
mathematician with Euler's abilities. (There is, of course,
no counterexample in third powers, since a counterexample in third powers would necessarily also be a counterexample to Fermat's famous "last theorem.")

The impact of computers on the center of pure
mathematics itself—on the construction of proofs for
theorems—is going to be even more important for
mathematics than the development of experimental
mathematical techniques. So far that impact has been
relatively modest. It will become less modest with the
advent of a generation of mathematicians who have
been familiar with computers from an early age.

A hint of the future impact of computers can be
found in the proof of the four-color theorem. A graduate student named Francis Guthrie apparently stumbled
on this problem when he asked whether four colors are
sufficient to color any map. In other words, he asked
whether it is possible to divide up a Euclidean plane into
areas that could not be colored with four colors, such
that no areas having a common border have the same
color. Mathematicians as distinguished as De Morgan
and Hamilton tried to prove this theorem without success. The problem became infamous as "the most easily
stated, but most difficult to answer, open question in
mathematics" (Stewart 1996, 105). Oddly, the problem
was solved first for complicated surfaces such as a torus
(doughnut-shaped surface), where seven colors were
found to suffice. The proof of the four-color theorem

on the plane was finally attained by Haken and Appel in 1976 using 1,200 hours of time on a then-powerful computer. The novel thing about Haken and Appel's proof is that a human mathematician has no way to check the computer's work. It would take too many life-times.

An even more stunning application of computer techniques to mathematical proofs was reported recently in an article by John Casti in *Nature* (1997). Casti describes a computer program written by William McCune at the Argonne National Laboratory. McCune's program has actually constructed a proof of the Robbins conjecture, a problem in Boolean algebra. The proof of the Robbins conjecture has eluded experts for more than half a century, despite work by mathematicians as distinguished as Alfred Tarski (whose Banach-Tarski paradox may be the strangest result in modern mathematics. See the discussion in Stewart 1992, 173–76). Actually, McCune's program found more than one proof of the conjecture. Casti notes that McCune's approach is completely different in its fundamental concept from the one used by Haken and Appel. In the four-color case, the computer merely searched through an astronomically large but finite number of possible maps that were specified in advance by Haken and Appel. In effect, the computer searched exhaustively for a counterexample. But McCune's computer program actually constructed the proof from scratch. Further, McCune's proof, after it was found by the computer, was simple enough to be checked by hand by experts in the field. According to Casti the project supervisor at Argonne, Larry Wos, has predicted, "This result may mark the beginning of a new era in mathematics, one in which mathematicians focus on producing interesting conjectures, leaving their proof or disproof to comput-er programs." Such a development would change mathematics beyond the recognition of any mathematician of the last few millennia.

These computer-aided and even computer-constructed proofs may be unusual today, but in the long run it is the other kind of proof, the proof that can be constructed without computer aid, that will become the oddball. Eventually most proofs will require some kind of computer assist. The plausibility argument for this is both straightforward and persuasive.

If we begin by considering all possible proofs in mathematics, we can place them in rough order of increasing complexity, for example, by counting the total number of printed pages in the shortest proof. At some point in this series of proofs they become too complex to be devised by unaided human mathematicians. Call this point P1. P1 is probably located somewhere just beyond the proof that there exist exactly twenty-six sporadic finite simple groups, which occupies about 10,000 pages (Stewart 1992, 125–27). Then there is a second point in the series at which the proofs become too difficult to devise even with computer aid. Call this point P2. (The location of P2 will shift as computer technology advances.) I am simply making the reasonable conjecture that the number of proofs in the interval from the beginning to P1 is far smaller than the number contained in the interval between P1 and P2. In fact, it seems likely that the ratio of the numbers is similar to the ratio of human to computer calculating speeds, within a few orders of magnitude.

The enormous capacity of computers to perform logical operations rapidly and accurately, and to check myriads of possibilities (as in the four-color proof) renders this conjecture almost irrefutable, even though it is not strictly provable. The proof will come by demonstration over the coming decades when the sheer number of computer-aided proofs begins to overwhelm the number of conventional proofs. The proofs that exist in the interval between P1 and P2 will be simply too intrinsically interesting to leave alone.

One might argue that all proofs are uninteresting beyond some point on the list (call it P3). This argument recalls a familiar conundrum that asks whether any uninteresting numbers exist (i.e., uninteresting positive integers). The difficulty is that if uninteresting numbers exist, then there must be a smallest one, and the smallest uninteresting number is, ipso facto, interesting. In point of fact, there is no evidence at all that proofs become less interesting or important as they become more complex. Rather, the existence of an extremely important proof at the 10,000-page level strongly suggests that they do not.

Mathematicians have raised considerable resistance to the concept of computer-aided proofs. Davis and Hersh, discussing the four-color proof, remark, "So it just goes to show, it wasn't a good problem in the first place" (Stewart 1992, 153). This resistance echoes the constructivist position in mathematics. The constructivists were mathematicians who refused to accept proofs of the existence of something in the absence of a formal method for constructing that something. For example, Cantor produced a famous proof that transcendental numbers exist, but the proof fails to show how to construct a single example of either a transcendental or a nontranscendental number (Stewart 1992, 293). The constructivist position draws considerable sympathy: Everyone would like to see all existence proofs made demonstrative. But in too many important cases we simply do not know how to produce the necessary construction; in the meantime, the constructivist approach misses too many important results.

A similar division will probably develop in the future between mathematicians who accept proofs that require computer work that cannot be checked "by hand" and those who do not. But in the end mathematicians will accept computer-assisted proofs when they turn out to be interesting and important and unattainable by other methods. After all, mathematicians are just like the rest of us: They lust after power. And computer techniques represent a level of power that they will not be able to resist for long. The mathematicians who use computer techniques will have too much fun, and the rest will sooner or later want in on the act. Mathematicians of the future may wonder that it was ever possible to work without computers, just as it is reasonable to wonder that it was ever possible to work without the axiomatic framework introduced by Euclid.

The resistance to the introduction of computers to pure mathematics may have another parallel in the resistance of some to the introduction of calculus itself. Intellects of the caliber of Bishop Berkeley correctly objected to the somewhat cavalier use of infinitesimals, which were not rigorously defined for a century or so after Berkeley's time (Stewart 1992, 102–3). But eventually the difficulties were worked out, and the appropriate rigorous methods were developed. Similar rigorous techniques will have to be developed for the validation of computer-aided proofs. Eventually, computer techniques will be recognized to be just another set of tools, analogous to techniques such as Fourier analysis or group theory. But these tools have extraordinary and enormous power, not to mention breadth of application.

These technical and conceptual revolutions have naturally caused many to think that science and mathematics have achieved a state of great sophistication, as indeed they have. Some have even argued that physics is near completion (Lederman 1993; Weinberg 1992), and one recent book claims that all of science is about to end (Horgan 1996). But this is exactly what both the

Pythagoreans and the Newtonians thought; yet today science and mathematics have taken off in directions that would be strange and unfamiliar to them. And with the computer revolution they will take off in still different directions.

In chapter 1 I argued that the invention of computers marks the beginning of a new level of civilization. Here the implication is that computers also mark the beginning of new levels for both science and mathematics. Outstanding work has been done in the past, just as the invention of calculus was preceded by the work of mathematicians of the caliber of Euclid, Archimedes, and Fermat. But Newton's invention of calculus was so important that it changed mathematics beyond all recognition. Similarly, computers will change science and mathematics beyond all recognition. The solutions to problems that have baffled investigators for decades or even centuries will be reduced to elementary exercises for students. Science and mathematics will be so different that it is no exaggeration to say that what they will become began here, with the invention of computers. Both science and mathematics are near their beginning, not their end.

4

Uncomputable Numbers

In the previous chapter I argued that computerized mathematical methods and techniques are about to change mathematics beyond all recognition. Some people may be uncomfortable with the idea that mathematics could be changed or even make any progress at all. After all, mathematics is frequently taught in schools as a closed body of doctrine, a finite set of rules for attaining "correct" answers and for establishing absolute, irrefutable truths. Indeed, prior to the computer revolution this was the conventional opinion of many mathematicians. Students who are saddled with such a misconception often fail to discover that mathematics is the purest and most exacting of the various forms of art, requiring creativity of the highest order.

To help dispel this misconception this chapter gives a brief sketch of some of the strange and counterintuitive discoveries that have already been made in mathematics during the computer revolution. These should give the reader some indication of the type of breakthroughs that might be expected in the next level of civilization.

As we saw in chapter 2, the mathematical revolution that accompanied the development of computer technology has provided fundamental insights into the nature and limitations of mathematics itself. But of all of the discoveries to come out of this revolution in mathematics, perhaps none is stranger or more disturbing than the discovery of uncomputable numbers.

As with so much of the computer revolution, this surprising discovery sprang from the fertile intellect of Alan Turing. Turing's work was based on that of David Hilbert and Kurt Gödel, which in turn was based on the invention of set theory and transfinite algebras by Georg Cantor. One of Turing's profound insights led him to ask the startling question: Is there a number that cannot be expressed by any mathematical formula at all? He called this an "uncomputable" number, and he realized that the question is equivalent to asking whether there is a number that cannot be calculated by a computer program.

Turing's question seems at first to be paradoxical, even nonsensical. How could there be such a thing as an uncomputable number? It would be beyond mathematics, in the sense that mathematics itself would not be powerful enough to express such a number.

The key to understanding the significance of Turing's question lies in an understanding of the similar crisis that was touched off by the discovery of irrational numbers by the Pythagoreans in about 500 B.C. The Pythagoreans knew, of course, about the positive integers (1, 2, 3,...), and they knew how to add, subtract, multiply, and divide integers, although lacking place-value notation for numbers, their techniques for multiplication and especially division were incredibly tedious and clumsy. Awkward as they were, the Pythagoreans' techniques were good enough to allow them to handle the arithmetic of the rational numbers as well as integers (a rational number or fraction is defined as a ratio of two integers, i.e., 3/5, 10/3, etc.). And the Pythagoreans knew a great deal about the properties of integers and rational numbers. They knew about prime numbers, for example, and Euclid was to prove that the number of primes is infinite. (A description of Euclid's proof is given in box 1, page 85.) Further, they knew that the rationals are dense. In other words, given any two distinct rational numbers, no matter how close they are in magnitude, there is an infinite number of rational numbers between them. The argument is straightforward: The mean or midpoint between any two rational numbers is rational. And the midpoint between one of the numbers and the original midpoint is rational, and so on. (See the discussion in Courant and Robbins 1996, 57–58.)

If an infinite number of rational numbers are contained in any interval, no matter how small that interval is, then that would seem, naively, to be enough numbers. Why on earth would anyone ever need any more than that? And yet the Pythagoreans were able to prove that the square root of 2 is not a rational number, that there are no integers a and b whose ratio gives the square root of 2. A modern version of the proof is given in box 2, page 86. According to some accounts this discovery so frightened the Pythagoreans that the mathematician who discovered irrational numbers, Hippasus, was taken out in a ship and thrown overboard (Dunham 1990, 9–10). Today they might simply cut off his research grant. But the historical records are a bit vague: Rather than discovering irrational numbers, Hippasus's crime may have been revealing their existence to nonmembers of the secretive Pythagorean sect.

The discovery of irrational numbers gave the first clue that arithmetic was vastly more complicated and more mysterious than it appeared. For the Pythagoreans, an irrational number was something that could not be expressed within the mathematics that they knew. They had no idea how to perform arithmetic with such numbers or even how to write them down.

They simply could not understand how irrational numbers could exist. The uncomputable numbers present similar philosophical difficulties for modern mathematics, but at a far deeper level.

In the eighteenth century, the Swiss mathematician Leonhard Euler noticed that the existence of irrational numbers raised some very profound questions. The rational numbers can be defined as the numbers that satisfy linear equations. In other words, if a and b are integers, then the linear equation

$$ax + b = 0$$

defines a rational number, x, and all rational numbers can be expressed in this form. Since the existence of irrational numbers proves the existence of numbers that cannot be expressed with linear equations, Euler wondered if perhaps there exist numbers that cannot be expressed with an equation of any order. In other words, Euler asked if there is a number x that cannot be expressed as the solution to an equation of the form

$$a + bx + cx^2 + dx^3 + ex^4 + \ldots = 0$$

in which all of the coefficients $(a, b, c, \ldots)$ are integers. Euler called such a number a transcendental number. After more than a century of work, Hermite was able to prove that e (the base of natural logarithms, 2.71828...) was a transcendental number, and Lindemann later showed that π was also transcendental (Stewart 1996, 66).

Turing's question—whether there exists a number that cannot be expressed by any mathematical expression at all—is a natural generalization of Euler's question, and yet it is far more profound. The irrational numbers demonstrate the existence of limits on the power of linear equations. The transcendental numbers demonstrate the existence of limits on the power of polynomial expressions. Turing's question suggests the existence of limits on the power of mathematics itself. Turing was able to prove the surprising result that the uncomputable numbers do exist. Perhaps even more surprising, nearly all numbers turn out to be uncomputable. The computable numbers, including all of the numbers that are actually used in any mathematical calculation (such as 2, 19, π, and the square root of 2) are the genuine oddballs in the real number system.

Turing's proof uses an argument very similar to Cantor's diagonal proof, which we will examine later in this chapter (see Turing 1965). The proof centers on the fact that the computable numbers form a countable set, whereas (as we shall soon see) the real numbers do not. This demonstration that nearly all real numbers are beyond the power of mathematics itself is an insight as deep and troubling for us as the existence of irrational numbers was for the Pythagoreans. As we noted, rational numbers contained all of the numbers that the Pythagoreans knew how to express and manipulate. To them the irrational numbers were somehow not numbers. Yet from our vantage point today, we can see a long history of this type of discovery, the successive inventions of new kinds of numbers that could not be manipulated or even expressed using the rules that worked perfectly well for all of the numbers that were known previously.

The first such discovery was the rationals themselves. The first use of rational numbers is lost in the mists of history. In Egypt, at the time of the Rhind Papyrus (about 1700 B.C.) scribes were already handling rational numbers (Newman 1956). Yet the rationals do not behave the same way that integers do: To manipulate rationals you need a new algebra. In modern notation, the algebra of rational numbers uses rules that look something like this (cf. Courant and Robbins 1996, 52–54):

$$\frac{a}{b} + \frac{c}{d} = \frac{ad + bc}{bd}$$

Which in fact is exactly the way that the integers behave (just set $b = d = 1$), although the arithmetic of integers is not usually expressed in this fashion. This is typical of the invention of new types of numbers. New rules are developed that work perfectly well for both the old numbers and the new ones. The new rules replace the old rules, which apply only to the older types of numbers.

Similarly, the irrational numbers cannot be handled with the rules used to calculate the rational numbers. To deal rigorously with irrationals you need new notations and mathematical techniques including such things as nonterminating decimal fractions and infinite series, which can handle both rational and irrational numbers (Courant and Robbins 1996, 58–72).

But even before the difficulties with irrational numbers were sorted out, mathematicians encountered another type of number that did not behave according to the rules they knew: negative numbers. The development of negative numbers created a similar crisis. As Stewart noted:

> In the mid-1600s Antoine Arnauld argued that
> the proposition -1 : 1 = 1 : -1 must be non-
> sense: 'How can a smaller be to a greater as a
> greater is to a smaller.'... In 1712 we find
> Leibniz agreeing that Arnauld had a point.
> (1996, 155)

It turns out, as we might have expected, that there is no real problem here. What is needed is a new algebra, a new set of rules for handling numbers. The new rules look something like this (Courant and Robbins 1996, 54–55):

$$(-a)(-b) = ab$$

But mathematicians had no sooner begun to develop the algebra of negative numbers when another new type of number began to appear in their calculations: square roots of negative numbers. And again the cry was heard: These things are not numbers. They do not follow the rules that we use to represent and to manipulate numbers. And once again the cure was the same: New rules were needed, new algebras that could handle these numbers correctly. A sample of the new rules is (Courant and Robbins 1996, 88–92):

$$(a + bi)(c + di) = (ac - bd) + (ad + bc)i$$

Again notice that, just as before, the new rules apply perfectly well to the old numbers (this time set $b = d = 0$).

Now we come to the crux of the problem with uncomputable numbers. At first blush they might seem to be just the next step in this sequence, a new type of number that present mathematics cannot cope with but that some brilliant new discovery will allow us to handle with ease and alacrity. But this is not the case. Today, thanks to the work of Cantor, Gödel, and Turing, we know far more about the limitations of mathematics itself than any earlier generation of mathematicians did. And it is clear that the problem with uncomputable numbers lies in those fundamental limitations of mathematics itself, rather than with any inadequacy of our current methods and techniques.

A full understanding of uncomputable numbers is deeply bound up with fundamental questions concerning the nature of infinity. Mathematicians over the centuries have argued that infinity cannot exist, that it is not a number. Why? Because infinity does not obey the same rules that other numbers do.

Does this sound familiar? By now it should. And it

does not mean that infinity does not exist, nor that it is not a number. In fact, infinity is a lot of numbers. As we shall see, there is more structure among the infinite (or transfinite) numbers than there is within the more conventional finite numbers. But the transfinite numbers need a new algebra, a different set of rules, just as the irrational numbers and the imaginary numbers needed new sets of rules in the past. The transfinite numbers actually proved to be somewhat more tractable than the uncomputable numbers, which are ordinary, finite real numbers. This should not be too surprising: After all, transfinite numbers were discovered first—before the turn of the twentieth century, in fact.

The development of the algebra of transfinite numbers was largely the creation of one man: Georg Cantor. Cantor devised at least five major ideas so brilliant and original, so simple, elegant, and powerful, that any one of them would have been sufficient to be the crowning event in the career of a great mathematician. The first of Cantor's major ideas was the basic realization that infinities are not absurd or impossible, but are actually numbers that needed only a new algebra to unlock their mysteries.

Intellectuals as far back as Galileo had argued that infinite numbers were absurd because, for example, every positive integer can be multiplied by itself to produce a perfect square, and every perfect square has an integer square root. Therefore there are exactly as many perfect squares as integers, although the perfect squares constitute only a small part (a "proper subset") of the set of integers. This violates a fundamental mathematical principle that goes back at least as far as Euclid's Common Notion 5: The whole is greater than the part.

Cantor developed the fundamental insight that this Euclidean principle applies only to finite numbers. Further, he realized that the property of having parts that are equal to the whole is the defining property of infinite numbers. It is the single attribute that is possessed by all infinite numbers, and by no finite number. This is another of Cantor's original ideas.

The next idea of Cantor's is hinted at in Galileo's argument. Cantor realized that in order to deal with the question of what it means to say that one transfinite number is different from another, he first had to deal with the question of what it means to say that a pair of transfinite numbers are equal to one another. To settle this problem he developed the notion of a set, which is simply a definable collection of objects or elements. A set may contain a finite or an infinite number of objects. The question Cantor then had to resolve is: Under what conditions can we say that two sets contain the same number of elements? If the sets each contain a finite number of elements, there is no difficulty: We simply count the elements in each set. But if the sets each contain an infinite number of elements then we need a more powerful idea than counting. To count the elements of an infinite set is simply not practical; it would take too much time.

Cantor's new idea was the following definition: Two sets of objects are said to contain the same number of elements if those elements can be put in one-to-one correspondence; that is, if every element in one set can be matched up with exactly one element in the other set, with no elements in either set left out of the matching. This essential idea is easier to understand than it is to state. To take a simple example, suppose you are feeding lunch to a small army. Just as everyone is seated, you suddenly realize that you do not know whether you have the correct number of forks. And it would take hours to count both people and forks. But you do not need to count. You simply ask everyone to pick up a fork. You then ask if there is anyone who does not have a fork. If everyone has a fork, you then ask whether any forks are left on any of the tables. If no forks are left, then you know that you have exactly the right number of forks, even though you do not know how many forks or people there are. You have demonstrated a one-to-one correspondence between people and forks, and that is all you need to know.

This idea of one-to-one correspondence is dazzling both for its simplicity, and for its stunning power. Cantor realized that it is powerful enough to count infinite numbers. It is the basis of the algebra of the transfinites. The idea is so simple that it may seem odd to spend so much time on it, but it is so powerful and so critically important that it must be clearly understood.

The idea of one-to-one correspondence is closely related to counting, which is simply a process of putting objects in one-to-one correspondence with positive integers. But it is much more powerful than counting because you do not actually have to do all the work. You can simply describe the correspondence rigorously. For example, in Galileo's argument, the set of positive integers can be put in one-to-one correspondence with the set of perfect squares. No one would argue that any integer or perfect square has been left out or forgotten from the correspondence. Did we forget that 28,472 is matched with 810,654,784? No, we did not forget it. It is right there on the list. Yet we were able to construct this entire matching convincingly in a very short amount of time, a few seconds even, far less time than would be needed to count all of the squares.

Now at last mathematicians were in possession of a tool that is powerful enough to investigate the nature and properties of the transfinite numbers. And in the hands of a genius this simple tool revealed the most surprising results.

Cantor began examining the sets that could be put in

one-to-one correspondence with the positive integers. They are called "countable" or "denumerable" sets, for the obvious reason. Infinite sets of positive numbers such as even numbers, squares, cubes, or primes were trivially countable. More surprisingly, Cantor found that the rational numbers are countable. It began to look as if every infinite set was countable.

We now come to the most surprising result of Cantor's remarkable career. He next showed that the real numbers (rationals plus irrationals) are not countable. In other words, an infinity exists that is larger than the number of positive integers. The proof is so simple that it is easy to describe in a few minutes. Indeed many accounts of it appear in the popular literature, dating back at least to Hahn (1956) and Gamow (1961). The critical proof begins by assuming that the real numbers are countable, and then uses this assumption to derive a very simple contradiction, thereby showing that they are not countable. The proof is sketched in box 3, page 88.

This discovery, that there are at least two infinite numbers that are unequal, would have been enough by itself to place Cantor in the first rank of mathematicians. But he did more. With another proof that is only slightly more complicated than the one cited above, he showed that there are more infinite numbers. There are lots of them.

To sketch out how this proof works, we need to define some terminology and describe some of the theory of sets. A *subset* of a set is defined as a set that is made up of elements selected from the original set. The subset may contain all of the elements of the original set, or it may contain no elements at all. (The set with no elements is called the null set or the empty set. This set, by definition, is a subset of every set.) A *proper subset* is defined as a subset that is not empty and does not contain all of the elements of the original set.

Now if the original set has n elements, there are 2^n possible subsets. The proof of this is straightforward: There are two possibilities for the first element in the set (in or out of the subset), two more for the second element, two for the third, and so on, so the total number of possibilities is $2 \times 2 \times 2 \ldots = 2^n$. Similarly, there are $2^n - 2$ proper subsets (all of the original subsets minus the empty set and the entire set).

The essence of Cantor's new proof involves showing that the elements of a set cannot be placed in one-to-one correspondence with all of the subsets of that set. For a set with a finite number of elements the proof is trivial because 2^n is larger than n for all finite n. But for a transfinite number of elements we have to make the argument with more care. The proof is sketched in box 4, page 91.

Since the number of subsets of a set must be larger than the number of elements in the set, each transfinite number can be used to generate a larger number by calculating the number of subsets of a set with that number of elements. Cantor called the first transfinite number (the infinity of the counting numbers) $\aleph_0$. It then follows that:

$$\aleph_1 = 2^{\aleph_0}$$
$$\aleph_2 = 2^{\aleph_1}$$
$$\aleph_3 = 2^{\aleph_2}$$

An excellent and readable discussion of the details of Cantor's proofs can be found in Dunham (1990, 245–80).

Cantor then asked one more question that finally defeated even his genius: Does this list contain all of the transfinite numbers? Or have we missed any? In particular, Cantor asked whether any transfinite number exists between $\aleph_0$ and $\aleph_1$. He believed that there was no such number, but he was unable to prove it. We now know that he failed to prove this conjecture because it is false. There are numbers between $\aleph_0$ and $\aleph_1$, although, in an odd twist, you can ignore them and still have a consistent mathematics, just as you can develop a consistent mathematics of the integers while ignoring the existence of all of the numbers that lie between any two consecutive integers.

But Cantor missed yet another set of transfinite numbers. In an uncanny echo of Euler's and Turing's questions, mathematicians asked whether there are infinite numbers that cannot be expressed in this fashion. And again the answer is yes: There is a set of transfinites that are too large to be expressed in terms of other transfinites. They are now known as the "inaccessible infinities." The inaccessible infinities turn out to be important in Lebesgue measure theory (Stewart 1996, 69).

Is this the end of the story? Do we finally have a complete understanding of numbers and of what constitutes a number? Of course not. The process never terminates. Just when we think that we understand numbers, they surprise us once again. Even today, mathematicians are exploring the properties of another new type of number, the "nonstandard" or "hyperreal" numbers, that hold the potential of revolutionizing the calculus and vastly simplifying it, according to the proponents of the idea. The hyperreal numbers were developed to allow a rigorous definition of infinitesimal numbers (Stewart 1996, 80–88).

We now return to the question that began this chapter, the existence of uncomputable numbers. Using Cantor's algebra of transfinite numbers, we can straightforwardly show that the computable numbers form a countable set. And we know that the set of real numbers (including the irrationals) contains a larger infinity than this. Therefore, as we said, nearly all numbers are uncomputable. There are simply too many real numbers, and there are not enough formulas to cover all of them, even if each formula is able to express a lot of computable numbers.

What does an uncomputable number look like? It is hard to describe explicitly because the description cannot be done in terms of mathematics. Yet we can easily give an operational definition for the construction of an uncomputable number. You simply toss a coin or roll some dice or use any other genuinely random process to determine each digit of the number. When you have done this an infinite number of times, you will have (with a probability of one) generated an uncomputable number. This may help to clarify the idea that uncomputable numbers comprise nearly all numbers: It simply means that if you select a number at random, its digits are very likely to be random. In point of fact, every digit of an uncomputable number need not be determined randomly. It is necessary only to have an infinite number of such digits. An uncomputable number could consist of all zeros, for example, except for every millionth or billionth digit, which is determined randomly.

Therefore, even though our ability to define and manipulate new types of numbers will go on forever, the vast ocean of uncomputable numbers will always remain forever beyond our reach. They are as remote and inaccessible as the final digits of π.

The discovery of uncomputable numbers demonstrates the existence of limits on the capabilities of computers, which are in fact limits on mathematics itself. No computer program can calculate the value of an uncomputable number. Nearly all numbers are therefore beyond the power of both computers and mathematics. Few more startling and unexpected discoveries have ever been made in mathematics. And it may be hard to imagine that even stranger discoveries await us in mathematics. But we have no reason to think that the process will end, or that the next discovery will be less strange than the last. As J.B.S. Haldane succinctly put it, the world is not only stranger than we imagine, it is stranger than we can imagine.

Box 1
EUCLID'S PROOF THAT THE NUMBER OF PRIME NUMBERS IS NOT FINITE.

This proof by Euclid, which is one of the earliest proofs in number theory to survive, has the identical form of the other three proofs contained in this chapter. In order to prove that something does not exist, you first assume that it does exist (assume the converse), and then use this assumption to derive a contradiction.

Following the discussion in Courant and Robbins (1996, 22–23), to prove that the number of prime numbers is not finite, we first assume that there is only a finite number, n, of them. If n is finite, then in principle we could construct a complete list of all of the primes $P_1, P_2, P_3, P_4, \ldots P_n$ using an algorithm like the sieve of Eratosthenes (see Courant and Robbins 1996, 25). But we can then use this list to generate new prime numbers that are not on the list. To do this we multiply together all of the primes on our (complete) list and add 1, to generate the number A.

$$A = P_1 \times P_2 \times P_3 \times P_4 \ldots \times P_n + 1$$

The critical fact about A is that when you divide it by any of the prime numbers on the list of primes you get a remainder of 1. Thus A is not divisible by any of those prime numbers. Now if A is prime then the proof is completed, because you have shown that there is a prime that is not on the list, which was complete under the assumption that the number of primes is finite. However, if A is not prime you are no better off, because A must then be divisible by at least two primes, and you already know that it is not divisible by any of the primes on your list. Therefore, again, there must exist prime numbers that are not on your list.

This is the essential contradiction that Euclid found: If you assume that the number of prime numbers is finite, then you can list them all and then use that list to generate prime numbers that are not on the list. Thus any finite list must be incomplete, and the complete list of primes cannot be finite. The structure of this proof is strikingly similar to Cantor's proof in box 3.

Box 2
PROOF THAT THE SQUARE ROOT OF 2 IS NOT A RATIONAL NUMBER.

Unfortunately, we do not know the exact form of the proof derived by the Pythagoreans; no text of the original proof survives. What follows here is a modern proof, but it is probably similar to the arguments made by the Pythagoreans. It does not use any concepts that would have been unfamiliar to them, and it is consonant with Aristotle's comment that the proof showed that a number must be both odd and even at the same time (Kline 1980, 104–5).

To prove that $\sqrt{2}$ is not rational, we once again assume the converse, that it is rational. In other words, we assume that there exist two integers a and b such that:

$$\sqrt{2} = a/b \qquad (1)$$

or

$$\sqrt{2}\, b = a \qquad (2)$$

Without loss of generality, we can assume that the fraction in equation 1 is reduced, in other words that a and b contain no common factor. This is critical to the proof. (If a and b do contain a common factor, we can always remove it by reducing the fraction to its lowest terms and then proceed with the argument.) Because we are dealing with exclusively positive numbers no problems are introduced by squaring both sides of equation 2:

$$2\, b^2 = a^2 \qquad (3)$$

And immediately we are in trouble, because no perfect square is twice the size of another perfect square. To see why this is so, let us complete the proof. We need to realize that the square of an even number is even, and the square of an odd number is odd (the proof of this will be left for the interested reader). Now a^2 must be an even number, because of the factor of 2 on the left-hand side of this equation. Therefore a must be an even number. We can then write:

$$a = 2\, c \qquad (4)$$

where c is the integer that is half of a. But substituting (4) into (3) gives:

$$2\, b^2 = (2\, c)^2 \qquad (5)$$

or

$$b^2 = 2\, c^2 \qquad (6)$$

But by the same argument this equation implies that b is an even number. Therefore both a and b must be even, and this contradicts our original assumption that *a* and *b* have no common factor.

There is another way to approach the problem that may be a little simpler and clearer, but it makes use of the fact that any integer can be written as a product of prime numbers in exactly one way, which had not been proved in Pythagoras's day but was proved by Euclid (proposition IX.14). The key to this proof lies in the fact that the prime factors of a perfect square always occur an even number of times (because each is repeated in the process of squaring—the details are left for the interested reader). Thus if you examine the prime factors of the numbers on each side of equation 3, you find that there must be an even number of 2s on the right side, but an odd number on the left (because of the extra 2). And this is impossible because the prime factorization of a number is unique. This is an example of proof by parity check, or a check on oddness and evenness.

Box 3
CANTOR'S PROOF THAT THE REAL NUMBERS (RATIONALS PLUS IRRATIONALS) ARE NOT COUNTABLE.

Cantor's proof is startlingly similar to Euclid's proof in box 1. Euclid assumed that he could construct a complete and finite list of prime numbers and then used the list to generate a prime number that is not on the list. Here we will assume that we have a complete and countable list of the real numbers, and use the list to generate a real number that is not on the list.

Cantor's proof therefore begins as before by assuming the converse, that the real numbers are countable. And for convenience we will restrict ourselves to the real numbers between 0 and 1. (If they are not countable, then the entire set is not countable either.) If these real numbers are countable then they can be placed in one-to-one correspondence with the positive integers. The correspondence would have to look something like this, with the integers on the left and real numbers on the right:

1	.**1** 9 6 4 4 2 8 8 1 0 9 7 5 6 6 5 8 ...
2	.0 **3** 4 8 6 1 0 4 5 4 3 2 6 6 4 8 6 ...
3	.0 9 **6** 2 8 2 9 2 5 4 0 9 1 7 1 5 2 ...
4	.1 6 0 **9** 4 3 3 0 5 7 2 7 0 3 6 5 1 ...
5	.2 7 4 9 **5** 6 7 3 5 1 8 8 5 7 5 2 6 ...
6	.4 9 4 6 3 **9** 5 2 2 4 7 3 7 1 9 0 9 ...
7	.9 4 0 5 1 3 **2** 0 0 0 5 6 8 1 2 7 6 ...
8	.1 2 2 4 9 5 3 **4** 3 0 1 4 6 5 4 9 0 ...
9	.4 4 1 8 1 5 9 8 **1** 3 6 2 9 7 7 4 3 ...
10	.6 0 9 6 3 1 8 5 9 **5** 0 2 4 4 5 9 1 ...

Every integer appears on the left-hand side of the list. And if this list exhibits a one-to-one correspondence between the integers and the real numbers, as assumed, then every real number (between zero and one) must appear somewhere on the right-hand side.

We know what the left-hand side of the list looks like. And although we don't know what the right-hand side looks like, we can use it anyway to derive the desired contradiction. Whatever real numbers appear on the right-hand side, we can always construct a real number whose successive digits consist of the first digit from the first number on the list, the second digit from the second, and so forth, selecting the digits along the diagonal of the list. These digits are printed in bold type in the sample list above. (This is why the proof is referred to as Cantor's "diagonal proof.") In the case of this particular list, the number would look like:

.1369592415…

Next we construct a new real number by adding .111111111…to this number, neglecting the operation of carrying, that is, $9 + 1 = 0$. The resulting number is:

.2470603526…

Cantor then asked the deadly question: Where does this new number occur in our list? If the list contains all real numbers then this number must be there somewhere. But it is not the first number on the list, because it is different from that number in the first digit. It is not the second number because it is different from that number in the second digit. Similarly, it is not the third on the list, or the fourth, or the fifth, or…; it is simply not anywhere on the list, which we assumed was complete. And this argument holds no matter how the numbers are arranged on the right-hand side of the list. This is the contradiction that Cantor discovered, and it completes the proof.

There are a few more technicalities that Cantor had to deal with. He used a more rigorous method of generating the number that is not on the list, for example. But this is the essence of the proof: No matter how you try to set up a one-to-one correspondence between the positive integers and the real numbers, there is always at least one real number that is not on the list. We can *always* construct such a number. In fact, we can always construct lots of them. Thus no one-to-one correspondence exists between the positive integers and the real numbers. These two infinite numbers (the number of integers and the number of real numbers) are not equal, because the necessary one-to-one correspondence does not exist.

> **Box 4**
>
> Cantor's proof that the elements of a set cannot be placed in one-to-one correspondence with the subsets of that set:

Cantor's proof begins the same way as all of the proofs in this chapter: It assumes that the subsets of a set can be placed in one-to-one correspondence with the elements of a set, and it then uses this assumption to derive a contradiction, thereby showing that at least one subset is always missing from the matchup.

The argument from this point on is more subtle and requires a little more effort than in the diagonal proof (box 3). Basically, Cantor notes that when you match each element of a set with some subset of that set, there are two possible types of matching. In the first type, the matched element is a member of the subset that it is matched with; in the second type, the element is not a member of its matched subset. All of the matchings must be of one type or the other. If the elements in the set are labeled a, b, c,...(continuing with an infinite alphabet) then the matching might look something like this:

$$a - \{a,c,d\}$$
$$b - \{b\}$$
$$c - \{b,c,d,p,q,z\}$$
$$d - \{b,c,d\}$$

$$\cdot$$
$$\cdot$$
$$\cdot$$

$$p - \{b,e,q,z\}$$
$$q - \{a\}$$

r - $\{x,y,z\}$
s - $\{c,d,e,f,g,h,i\}$
.
.
.

Here the brackets denote a subset, and the dashed line separates the two types of matchings. The key to the argument is that the elements on the left side that are below the dashed line constitute a subset of the original set $\{p,q,r,s,\ldots\}$. Again, Cantor asks the deadly question: Where does this subset appear on the right-hand side of the list? In fact, it cannot be found anywhere on the right-hand side of the list. It cannot be above the dashed line, because each of the subsets there contains the element that it is matched with, none of which is contained in the critical subset. Similarly, it cannot be below the dashed line, because each of those subsets does not contain the corresponding element on the left-hand side, and all of those elements are in the critical subset. Therefore, no matter how you try to match the elements of a set with the subsets of the same set, you can always construct at least one subset that is missing from the matchup.

5

The Computer Revolution
in Education

The previous chapters have outlined the theoretical and conceptual revolutions that underlie the computer revolution. These are among the most profound conceptual revolutions in all history, far larger than the comparable revolutions that triggered the Renaissance. But the practical revolutions that will accompany the computer revolution are even greater than the theoretical ones. They will transform virtually everyone's everyday life.

The transformation of education may be the most important of the many practical revolutions sparked by computer technology. No facet of civilization will be altered more radically. Just as computers are about to replace books as our main repository of information, computers will come to occupy the central position in education once occupied by books. Computers will also take on many of the functions previously performed by instructors. More important, computer technology will alter the very goals of education.

To understand the changes that computers will bring about in the educational system we first need to examine the objectives and methods of the present system. Then we need to develop methods that can exploit the immense power of computer technology to accomplish not merely the present objectives but radically improved objectives.

Of course in any criticism of the present education system, it is important not to lose sight of its extraordinary accomplishments. The current system of universal education is an achievement that is unmatched throughout most of history. The civilization that invented the airplane, the electric light, the microchip, that walked on the moon, that made democracy work on a continental scale, can still count universal education as perhaps its greatest triumph.

But despite its unparalleled achievements, our present educational system is inadequate to meet the needs of today's society, to say nothing of the next decade or the next century. The problem is that the educational needs of society are growing at an explosive rate. We are no longer a society that can sustain large numbers of poorly educated laborers in farms or factories. The demand for minimally educated, unskilled labor is vanishing rapidly. Our schools, which for generations have been geared to train only a small elite for tasks that require creative intelligence, will soon have to prepare everyone for such tasks.

We may reasonably ask whether the entire population is capable of functioning effectively in creative fields such as science, mathematics, engineering, literature, philosophy, or the arts. We do not know the answer to this question, yet the answer is in some sense irrelevant. Our objective should always be to let everyone develop his or her abilities and creative potential as far as possible. I believe that we shall be pleasantly surprised to discover how high that potential is when it is fully tapped. By far the worst failing of our present educational system is that it develops only an insignificant fraction of the abilities of most individuals.

Make no mistake: *This* is the great challenge of the age. For an age that has reveled in spectacular challenges and achievements, nothing surpasses this one.

Meeting this challenge will not be easy. The problem is daunting, but it is not beyond hope. The main reason to be hopeful is that the very technology that is causing the problem is the same technology that can provide a remedy. We should unleash the power of computers to create an education system whose goals are far beyond anything that has ever been attempted before. But to do this we must be prepared to examine and challenge virtually every aspect of the system. We need to discard ideas that were merely adequate in our grandparents' time and to seek out novel and imaginative solutions to the problems that we face.

To meet the goal of fully educating every student we will need a fundamental change of attitude and perspective. The current system focuses on how much students learn. The focus instead ought to be on how much they *want* to learn. As soon as students want to learn, most of the problems that plague the present education system evaporate: Students who genuinely want to learn will work hard to overcome any obstacles they find in their paths.

Inducing students to want to learn might appear to be an impossible task. Our current public education system is so universally disliked by students that the entire education process appears to be an inherently unpleasant or painful experience. But this is pernicious nonsense. It overlooks the fundamental fact that curiosity is one of the most basic and powerful of human drives. An educational system should tap this marvelous resource and stimulate the innate curiosity of each student, stimulate it to the point at which the drive and desire to learn will come from within. That it largely fails to do so today is one of the worst faults of the present system.

Plato understood the problem. In *The Republic*, he wrote:

> Knowledge which is acquired under compul-
> sion obtains no hold upon the mind....
> [Therefore] do not use compulsion, but let
> early education be a sort of amusement. (1901,
> 234)

In a word, education should come to be dominated by the very function that children naturally use to educate themselves: play.

There is a good reason why the goal of making students want to learn through the use of play has not been widely achieved in the past: In the absence of computer technology it was simply too expensive. Creating an education system in which a student can play and still make steady and carefully monitored progress will require an immense expansion in the effort and resources expended on each student. Without computer technology there would be only one way to accomplish this: The instructor/student ratio would have to be radically increased, preferably to something in the range of about ten to one (i.e., ten instructors per student). Clearly this is not a reasonable goal. But the enormous power now available with computer technology offers the promise of providing the needed expansion of effort and resources at very reasonable costs. This possibility provides us with an alternative that is not just practical but imperative: We must use computer technology to increase the productivity of instructors, and increase it by orders of magnitude. Many of the routine teaching tasks that now consume a large portion of an instructor's time must be relegated to machines. The instructors will then be freed to focus their attention on the creative aspects of the education process. This solution is inexpensive, and its cost is dropping rapidly as technology advances.

To understand how computers should be used to

increase the productivity of instructors, we need to recognize that the learning process has two fundamentally different elements. The first element is the mastery of fundamental facts and skills, including such things as reading, grammar, vocabulary, history, geography, basic mathematics, and logic. The second element is the development of creative skills such as composition, exposition, problem solving, artistic expression, or, more generally, learning how to think.

The computer will have its primary effect on the first element. Indeed, it should change that element altogether. In education systems today most of a student's time is taken up with the first element, mastery of fundamental facts and skills. And this must remain the case; creative skills are largely useless without a mastery of the fundamentals. But today most of the instructor's time is also taken up with this aspect of instruction, and there is no excuse for such an appalling waste of valuable time and talent in the computer age. We have a better way to handle these tasks: Computers should take over most of the work of training students in basic skills and knowledge. Using the incredible speed and computational power of computers, we should be able to allow the students to play while inducing them to learn essential skills and knowledge. At the same time the computers can carefully and continuously monitor and evaluate their progress.

One of the key concepts for the development of computer software to meet these goals is the instructional computer game. We know that computer games possess both of the essential elements needed for this process. First and most important, students enjoy playing them. Second, success in the games requires a variety of fundamental skills, ranging from eye-hand coordination, to basic knowledge, to problem-solving skills.

A great deal of creative genius will be required to

devise games that students enjoy playing and that at the same time require them to develop the right mix of fundamental skills. Seymour Papert at MIT has developed one of the best examples of such an instructional game (Papert 1980). His LOGO program lets the student create pictures by directing the motions of an animated paintbrush whimsically referred to as a "turtle." By learning how to command the turtle to move and turn, the student approaches naturally many of the concepts required for algebra, analytic geometry, and computer programming. LOGO exploits the almost universal human desire to create pictures and to manipulate them. Building on this simple idea, Papert has developed an instructional program that leads young students with ease and even enthusiasm through bodies of knowledge that their elders, in more conventional instructional environments, often find difficult, tedious, and largely irrelevant to their interests.

The range of games that have instructional possibilities is limitless. Computerized simulations of real-life situations offer a wealth of possibilities. Business concepts and skills could be taught through simulation games. Engineering problems could be simulated, and the solutions, test bridges and buildings, could be displayed graphically and their faults and weaknesses exhibited through perhaps humorous simulated failures. Alan Kay (1991) describes computer simulation games that let third-graders practice designing a city, and another that helps students learn about animal behavior. And Dertouzos discusses examples of the use of simulators for things ranging from training airplane pilots to teaching executives how to handle management crises (1997, 180–82).

The essential element of computerized instruction is that the instructional games must be devised so that students enjoy playing them. *Why* they enjoy playing does not matter. It does not even matter how many of them enjoy it—a wide variety of different types of games will probably be required to suit the needs of students of varying interests, temperaments, and skills. It would be useful if the facts or skills being taught were a natural development of the game, such as using the context of orbiting spaceships to teach the fundamentals of Newtonian mechanics. The necessary information could also be grafted on more artificially, for example in a maze game that requires a variety of problem-solving techniques to get in or out. Whatever form the instruction takes, the important thing is that the student enjoy the experience, enough to *want* to spend time mastering it.

Much of the genius involved in the design of effective instructional software will lie in the development of patterns of reward and punishment used to motivate the students. Since the objective is for the students to enjoy using the software, the emphasis should be focused on rewards rather than punishment. But the rewards that students require are often remarkably modest. Simple approval by the instructor often works wonders. And students will work hard to simply improve a numerical score, if they enjoy the process. Humor and surprise are also effective reward techniques. But the best games will stimulate the student's curiosity, his or her desire to learn more about the subject being taught. This is by far the finest reward that could be offered, and when it is done skillfully the student whose curiosity is aroused will recognize and understand the value of this reward.

One of the fundamental failings of the present education system is that students commonly lack any clue as to the usefulness of the facts and skills they are expected to master. They generally have not the faintest idea why a knowledge of algebra or history would ever be useful to them, and they often pass their entire education experience without finding any such uses beyond their performance on justly hated examinations. Computer games can fill this void by providing environments in which the students want and need their newly learned facts or skills for their *own* purposes. It makes little difference whether the perceived need is similar to the needs that will be encountered in real life or is wholly artificial, so long as the student sees the need in the context of something he or she wants to accomplish. In fact, one of the most useful lessons to be learned is that basic mental skills and tools have a wide variety of applications in different games as well as in real life. The students who grasp this simple fact are ready to use their new skills away from the games and outside of the classrooms.

There is another aspect of computerized instruction that may have particularly dramatic effects on the education process. Instructional game programs should be able to analyze and monitor a student's ability and progress far better than can be done with conventional examinations. Examinations are such a traditional feature of education that it may come as something of a shock to realize that they are already technologically obsolete. In the pre-computer era, examinations were unavoidable—there was essentially no other way to measure and keep track of a student's achievements and progress. But examinations never did this job very well, and with computer technology there are novel techniques that can do it far better.

What is wrong with examinations? In the first place, they require so much effort on the part of both the student and the instructor that they are not done often enough. Intervals of weeks to even months often pass without adequate assessment of the student's performance. Next, examinations are seldom able to adequately cover all of the instructional material. Chance frequently determines whether the questions that happen to be asked are the ones that were studied or crammed for. In addition, the results of a test do not adequately describe the actual performance of the student. The score on a test usually gives only the percentage of right/wrong answers, with no consideration for the reasons why some of the answers were wrong. For example, a student who does not know the correct answers may need a kind of remedial help different from that needed by a student who knows the answers but is unable to complete the test with adequate speed. But examinations seldom discriminate among these and other problems.

More seriously, examinations often engender fear, which can be extremely detrimental to the learning process. Fear of punishment (poor grades) is sometimes an appropriate tool for enforcing desired behavior, but it is a tool that should be used sparingly and restricted to rather serious infractions. Instead, fear often has the negative effect of making the students perform more poorly than their abilities warrant.

The reverse problem is sometimes encountered: Some students manage to master the art of taking examinations to the point that they are able to score higher than their capabilities warrant, particularly on specialized tests such as standardized multiple-choice tests.

Another serious problem with examinations is the lack of immediate feedback. Exam scores are often not known for days or even weeks after the exam. Then when they become known, and the student receives some (belated) incentive to perform better, he or she frequently lacks an immediate opportunity to translate that incentive into action.

Computer technology can eliminate *all* of these problems. While a student is using an educational computer game to master a set of facts or basic skills, the computer, with its massive number-crunching capability, will be able to monitor the student's performance continuously and in much greater detail than is possible with conventional examinations. A well-designed instructional computer program will do far more than just record the percentages of right or wrong answers. It can keep track of the frequency of errors on each item of knowledge or skill individually. It can also record response times for each item. It can even modify the instructional portion of the program to give extra attention to the areas in which the student is having difficulty, either in speed or accuracy, while at the same time being careful not to make the exercise too difficult or too discouraging. (This important idea will be dealt with more fully in a later section.) The program might even be able to monitor for subtle performance cues such as signs of fatigue or boredom or even illness and respond appropriately.

And the computer can always provide immediate feedback to the student. When the student performs well, he or she should know immediately. A well-designed game will give an immediate and appropriate reward. When performance falls short, the student will not only be informed of that but will also be apprised of the reasons for the difficulty, and then be given an immediate opportunity to improve.

Let me make it clear that in advocating the elimination of examinations, I am not suggesting any reduction in the testing and evaluation of a student's progress and ability. I am proposing exactly the reverse, that testing should be dramatically increased to the point at which it is essentially continuous and far more detailed than is possible with conventional examinations.

Computers can provide a novel way to evaluate the performance of the student by using the technique of interactive evaluation. With this technique the instructional program varies the level of difficulty of the problems offered to the student in a way that depends on the level of performance on previous problems and questions. The evaluation or "grade" then depends not so much on the proportion of right/wrong answers (a crude measuring technique) but rather on the level and type of answers that are given and the problems that are solved. This technique is essentially impossible without the colossal calculating speed available with computers. One advantage of this technique is that it tends to function much faster than conventional examinations because far less time is wasted on problems that are too easy or too hard. In addition, because the program focuses on exactly the right difficulty level for each student, the student often perceives the results to be fairer, a more accurate representation of his or her actual ability level.

In evaluating the student's performance and providing him or her with continuous and immediate feedback the computer is not doing anything that a human instructor could not do, given enough time. But there is no way to give him or her enough time. The amount of bookkeeping and arithmetic required for the task makes it essentially impossible for a human instructor to attempt it even for a single student, let alone for a class. It is not even feasible for the case where each student has many instructors. The level of calculating power that can be provided by computer-aided instruction thereby allows the development and application of radically new and effective teaching techniques. As is frequently the case with problems in science, engineering, and mathematics, the sheer power available with computers provides solutions to problems that *no* reasonable level of effort could provide without them.

The instructor's task in such an environment would

be to find the particular set of instructional programs best suited to each student's abilities and interests. Selecting the proper combination of games and programs could be as simple as trial and error, or it could be guided to some extent by aptitude testing and analysis. The instructor will be relieved of the tedium of producing, grading, and recording examinations, to say nothing of the loss of valuable time involved in the repeated presentation and review of routine material. Instead, the instructional material should be presented by the computer game and reviewed only in those areas in which each student demonstrates a need.

The time saved for the instructor should be used to focus on the needs of students who may require special attention, those who are having difficulties that the instructional software is unable to alleviate. By carefully monitoring the computer-generated scores, the instructor could render individual aid as needed. He or she might be able to suggest other instructional games that cover similar material in ways that might better appeal to the individual student. The instructor's role would become similar to that of a trainer or coach, rather than the present somewhat unpleasant roles of overseer, taskmaster, judge, and jury.

The students should find that developing basic skills is a pleasant and enjoyable process once the instruction is tailored to their individual needs and interests, and paced at a level that each can handle. If the process is paced correctly and provides steady reinforcement and reward, it should work not only to build the students' knowledge and skills but also to help them develop their self-confidence and self-esteem, making them eager to learn more and explore their new capabilities.

The development of the necessary instructional software will not be easy. As I said, it will require nothing less than the finest creative genius that the human race can muster. Fortunately, however, it will not require such genius often. Once per game will do nicely. And there are large reservoirs of talent that can be brought to bear to create this software, in part because almost anyone can attempt it: The capital investment required to successfully develop computer games is quite small, within the reach of most individuals, and almost insignificant on the scale of the budgets of even small colleges, education departments, and research institutions. This presents an unusual opportunity for education departments and even talented individual instructors to reach out and help very large numbers of students, many more than can be reached with conventional educational techniques. And once the software is developed, tested, and refined, it can be rapidly and inexpensively disseminated.

These instructional programs can be used at first outside of the usual curriculum as an auxiliary tool for supplementary and remedial work, until their value is proved. They can then be added to the curriculum gradually, as instructors and students discover and demonstrate their usefulness. This software can therefore be used experimentally without disrupting current education practices until its effectiveness is proven. This is a substantial advantage over previous experimental educational efforts such as the "new math" of the sixties and more recent efforts at non-phonic reading techniques, whose failures may have done serious and irreparable harm to some students.

If these instructional games are sufficiently attractive and enjoyable, their use could conceivably cut into the vast reservoir of time that students currently spend watching television and in other idle pursuits. One shudders to think how the television industry might respond to this threat to its audience and revenue base.

Another important advantage to developing a set of

effective instructional software is that the same tools that are effective in teaching young students could also be made available for continuing or adult education. If the games were sufficiently attractive they could effectively create an education process that would become a permanent "cradle to grave" activity for everyone. Also, as instructional software becomes individually tailored to the needs of each student, many of the problems associated with monolithic school edifices and bureaucracies could be reduced or eliminated. Home schooling could replace these anachronisms on a broad scale.

Some people might protest that if these instructional games become very popular and widespread in use, they could have the undesirable side effect of producing a generation of students who are facile in the use of computers but less able to interact with or relate effectively to their fellow students. This is indeed a danger, and one that is not new, but is currently somewhat rare. Some students today relate better to their books than to their fellow students. But if this problem were to become common, I can well imagine that many instructors would find it an improvement over their current difficulties.

The problem can be circumvented in several important ways. In the first place, with less of the instructor's attention devoted to basic drills, more time will be available to deal with problems related to students' interrelationships. Perhaps more important, with extensive use of computer communications, students should be able to interact with a far broader range of other students than can be found in a single classroom, school, or neighborhood. Regular meetings of various student interest groups could be arranged and organized with the aid of computer communications, and meetings could be held either on-line or face to face, whichever is feasible. The instructor's contribution to such activities would include keeping the students informed of the existence and performance of groups that relate to the subject of instruction or the interests of the students. The opportunities for the students to develop skills for personal relationships should be extended rather than restricted by the extensive use of computers, if it is handled correctly.

In addition to the development of basic skills, knowledge, self-confidence, and self-esteem, computers will be able to contribute to the more creative aspects of the learning process in many ways, some of which will not be easy to foresee. At the simplest level, the use of word processors and symbolic algebra software will relieve the student of much of the tedious labor that was once an unavoidable component of the creative process, and thereby allow far more time and effort to be spent on the genuinely creative aspects of composition, mathematics, and so on. The use of word processors complete with spelling and grammar checkers, and thesauruses opens a variety of new opportunities for more effective teaching of creative writing techniques. Papert (1980, 31) describes how instructional techniques for composition and creative writing can change in an environment where the production of a new draft is nearly effortless. Both the instructor and student will be able to focus much more of their time and attention on the development of effective and indispensable techniques such as revision, analysis, word choice, organization, and the other essential elements of the process of composition.

Programs such as Papert's LOGO were actually designed to teach creative mathematical skills rather than more basic techniques. Papert argues cogently that computer techniques will actually change the way we think, and lead to more effective ways of perceiving and understanding the world in which we live (Papert 1980, 19–37).

Yet another fundamental change to education will be

caused by the computer revolution: Computer technology is going to make radical changes in the mix of skills that students need to develop. Of the traditional "three R's" of education, only reading is going to survive unscathed. Indeed, the need for reading skills will be greatly enhanced. In the computer age people will have access to enormous amounts of information that they will need to absorb, assimilate, and act on; all of this will require reading (except in the unlikely event that we are able to develop the technology for really novel learning techniques such as direct nerve induction or oral ingestion).

The second R, writing, in the sense of handwriting or generating letters and words using pencil and paper, has been largely obsolete for several generations, at least since the invention of the typewriter. Students will always need some minimum level of ability to print letters by hand, but beyond that level keyboard skills will be far more important than penmanship. And as soon as the technology of computerized voice recognition is developed, even keyboard skills will become obsolete. Voice recognition already exists in rudimentary forms on a variety of computer platforms, and it will only get better in the future.

The third R, arithmetic, the use of pencil and paper to do sums and quotients, should be largely eliminated from the curriculum by the computer revolution. Students should be taught the effective use of calculators, spreadsheets, and symbolic algebra software as soon as they learn to read. There is no point in misusing education to perpetuate obsolete technologies, and the percentage of arithmetic done today with paper and pencil is essentially zero; indeed it is not significantly different from the percentage of arithmetic that is done today with Roman numerals.

Obsolete skills, such as the use of handwriting, manual calculation, Roman numerals, Egyptian fractions, slide rules, and clay tablets, should be pursued by those who have a historical interest in such methods and technologies, but should not clutter the minds (and take up valuable education time) of those whose interests lie in other directions. Pencil and paper calculations should therefore be restricted to consenting adults and should under no circumstances be permitted to minors. This may raise the obvious specter of creative but mischievous students sneaking into the public library to look at obsolete arithmetic texts, but such problems can be dealt with through careful attention and firm discipline.

In sharp contrast to the standard practice today, little or no rote memorization should exist in mathematics instruction. The emphasis of education in mathematics should instead be on developing problem-solving skills, learning to make effective use of the most efficient techniques and technology available. Students should be particularly trained in the use of the technologies that predominate outside of the classroom. We would do far better to train a generation that knows how to use computer technology to effectively solve a broad range of mathematical problems than a generation that has memorized addition and multiplication tables but has no idea how many fruit trees are contained in an orchard six rows by eight.

The current emphasis in elementary mathematics education on rote memorization of arithmetic skills is one of the prime contributors to the math-phobia that is endemic throughout our culture. And as Stewart noted, most students conceive of a mathematician as "...an earnest, bespectacled fellow poring over an endless arithmetical screed. A kind of super-accountant" (Stewart 1996, 2). If computer techniques can contribute to the elimination of these phobias and stereotypes, so much the better.

Some people have argued that learning arithmetic

with pencil and paper teaches familiarity with numbers. This, again, is pernicious nonsense. Familiarity with numbers can be taught just as well with a calculator or spreadsheet. This argument is as senseless as the argument that creative writing and composition can be done better with pencil and paper than with a word processor. Oddly, professional mathematicians have apparently been aware of this for some time. Stewart says that among the mathematicians of his acquaintance, "none of them is especially good at arithmetic" (Stewart 1996, 2). It has also been argued that children may develop an appreciation for the important concept of algorithms from the study of arithmetic. The difficulty with this argument is that arithmetic is one of the worst tools for studying algorithms. Computer programming provides a much more natural milieu for the teaching of algorithms, and Papert's LOGO program provides a perfect environment in which students *enjoy* learning about algorithms.

Determining the exact mix of skills that will be required in a technological society will never be an easy task, and the mix will change constantly. Technology changes so rapidly that yesterday's breakthrough invention is already obsolete. For example, pocket calculators are new enough that there is still some resistance to their introduction in the schools, and few instructors are aware that these calculators are already as obsolete as the sliderule or the typewriter. Except for the simplest calculations, or for situations in which portability is an overriding consideration, the calculator has been superseded by more powerful software, especially spreadsheets, which should be taught as "number processors." Indeed, the difference between a calculator and a spreadsheet is precisely analogous to the difference between a typewriter and a word processor. With a calculator, a single errant keystroke can ruin a long calculation. And when the calculation is finished, often the only way to check the result is by repeating it, keystroke for keystroke. With a spreadsheet the identical calculation can be done in a sequence of cell entries, and when the calculation is finished every keystroke is still displayed for checking. And, as with a word processor, whenever an error is found the entire calculation is instantly and effortlessly finished as soon as the error is corrected.

The pace of technological change today is so rapid that older technologies are continually being made obsolete, and the determination of which ones will be most useful for students throughout their lives is never easy and never constant. Yet instead of welcoming the new technologies, educators often resist changes to the curriculum. This problem is literally as old as the pyramids. Plato, in the *Phaedrus*, relates an Egyptian legend that describes how the god Thoth taught the first pharaoh the art of writing hieroglyphics. According to the legend, the pharaoh complained to Thoth that the invention of writing would ruin the minds of his scholars because they would no longer be required to memorize everything. We have long since forgotten the argument that students should not be taught reading and writing. Instead, these skills are now customarily taught at the very beginning of the education process. Similarly, effective computer skills need to be taught at the very earliest levels of instruction. We should give students the proper tools, develop their motivation, and then turn them loose. Set them free and then watch what they will accomplish. We shall be very proud of them.

Curiosity is one of the defining traits of human beings. It is such a powerful human drive that stimulating it, developing a student's motivation and desire to learn, ought to be no more difficult than stimulating an adolescent's interest in the opposite sex. And developing an education system that makes students want to learn is the fundamental key to constructing a society in which all individuals are able to exercise their creative intelligence to the fullest.

Of course no modification to our education system will be perfect; we will not be able to solve every student's problems. Humanity is too diverse to permit such totality. We should be satisfied to be able to solve many problems and find approaches that will appeal to large numbers of students. The range of appeal will gradually widen as more and more human genius is poured into the development of computerized instructional software. But the sole criterion for judging an education system should be the question of how much it makes students *want* to learn. How much they actually learn is a vastly less important question that will largely take care of itself when the students want to learn. The poet and playwright W.B. Yeats may have expressed this idea best with his apt metaphor: "An education is not the filling of a pail, it is the lighting of a fire." Computer technology gives us new and completely unprecedented methods for igniting that fire.

In previous chapters I have argued that the invention of the computer marks the beginning of a new level of civilization, as well as the beginning of a similar new level of science and mathematics. We should not be surprised to find that it also marks the beginning of a new level of education. Computers are such perfect tools for the education process that future generations will wonder how it was ever possible to have an education system without them, just as earlier generations may have wondered how it was ever possible to have an education system without printed books or without writing.

No one can foresee all the developments in computerized education. Many of them will no doubt come as a surprise to everyone. But the one thing that is perfectly clear is that a computerized education system will resemble current systems about as much as the current ones resemble medieval education systems. Computer technology is about to cause the largest change in education since the introduction of the printed book.

6

Language in the Computer Age

Many of the changes that computer technology is bring-
ing, such as those in engineering or finance, are clear
and obvious. But changes in other fields may be more
subtle and far less obvious; that does not mean they are
necessarily less profound, however. Spoken language,
for example, is one aspect of civilization that might
seem immune to the revolutions generated by this technological marvel. The impact of computer technology
on language will be far different from the corresponding
effect on science or on education, for example, because
in language there is no comparable imperative for
change. In fact, computer technology might have no
effect at all on language. So rather than asking, "What
effect will computer technology have?" we should
instead ask, "What effect would we like it to have?" If we
can identify changes in language that would be desirable, we could try to see how the new technology might
be used to expedite or promote these changes.

Languages have a number of problems that we might
want to correct, but one problem stands out over all the
others: There are far too many languages. The problems
caused by the plethora of languages include misunder-
standing, miscommunication, and absence of communication, to say nothing of the huge waste of time and

effort involved in translating between languages or the wasted effort of learning various languages in the first place. And there are few, if any, corresponding benefits that might mitigate all this trouble and waste. No demonstrable benefit results from having more than one language.

The desirability of a single universal language has been clear for centuries, and modern communication technology has made the need even more imperative. Yet none of the numerous attempts to develop a universal language has ever met with any particular success. Could computer technology be of help here? Is it possible to identify particular reasons for the failures that might be alleviated using the power of modern computers? A brief analysis of those failures seems to indicate an affirmative answer to these questions.

Three general methods have been suggested for the development of a universal language. The first method proposes the revival of a "dead" language such as Latin. The second involves the use of artificially constructed languages such as Volapük, Ido, or Esperanto. The third method advocates the universal adoption of an existing language, such as English or Chinese, that presently has a large number of users, both native speakers and those who speak it as a second language.

The key reason for the failure of the first two methods is that the number of people who are able to speak and understand the proposed language(s) has never been large enough. This is unlikely to change in the future. The problem of not having a large enough base of fluent users of the language is not unlike the problem of trying to introduce a new recording medium such as compact disks to replace vinyl LPs: It is not profitable to produce the recordings until enough people own the machines that can play them, and no one will buy the machines until recordings are available. Yet despite this chicken-and-egg problem, new recording media have been successfully introduced. Why should a new universal language not have similar success? Perhaps if a profit motive were involved the situation might improve, but how such a motive might come about is unclear. Indeed, there is a cost involved in any change.

The third method, universal adoption of an existing language, avoids the problem of not having enough people who can speak and understand the language. Unfortunately it encounters a different set of problems, not the least being nationalistic pride in those regions where some other language is spoken. But none of the possible candidate languages possesses all the features that might be desired in an ideal universal language. All of them have quirks and idiosyncrasies that have no actual function other than to make the language unnecessarily difficult to learn and use. The irregular and nonphonetic spelling in English is perhaps the most obvious example of such a problem.

A combination of methods two and three might provide a useful solution to all these problems. The idea would be to create a language that is similar enough to a current language that it would be immediately and easily understood by a very wide audience. At the same time, because the new language would have its flaws and idiosyncrasies removed, it would be substantially easier to learn than any existing language. It would be the ideal second language for *everyone*, especially those who are native speakers of the original root language. And once it became everyone's second language, it could rapidly acquire the desired universality of usage.

Although variations on this idea have been tried and have failed, computer technology might be able to alleviate some of the reasons for its past failures. Computer technology might in several ways provide the necessary push that could make a difference for the general acceptance of the new language. For example, the overriding problem with the English language is, as I noted, its idiosyncratic and nonphonetic orthography (spelling). Many attempts have been made to improve the spelling of English, including efforts by George Bernard Shaw and the *Chicago Tribune*. The only effort that had any appreciable effect was the relatively minor set of spelling revisions introduced by Noah Webster in the early part of the nineteenth century, and the use of Webster's spellings is largely confined to the United States.

But the introduction of computer technology could have a dramatic effect on English spelling, if we were to allow it to happen. The critical point will occur with the introduction of voice-actuated word processing software. This software will allow a computer to recognize spoken language, and thereby replace the keyboard as the principal input device for word processing. The obvious advantages of this technology are ease and speed. Virtually everyone can speak words several times faster than even the most skilled typist can type them. Primitive versions of the necessary hardware and software for voice recognition are already functional in laboratory settings, and the vagaries of English spelling are among the problems that impede the development of this technology. If a standard, regular, and phonetic English spelling were available it might facilitate the development of voice-actuated technology. Thus we may be facing a golden opportunity, a once-in-a-millennium chance to rid the English language of its most repugnant and universally disliked attribute. The lure of voice-actuated word processing may be great enough that the resistance to changing English spelling could be overcome at last.

If so, the window will be brief, and the critical time period is nearly upon us. Success or failure will be decided between the time that the voice-actuated technology is first commercially available and the time at which sufficient computer power is available to handle the present execrable orthography. The decisive period might be a matter of years, or it could be even less. If we are to take advantage of this opportunity a regular, phonetic spelling system must be developed and, more important, advertised and publicized enough to make it broadly acceptable; otherwise the voice-actuated computer systems will not be able to make any use of it. What use would a voice-actuated technology be if it produced output that nobody used?

Computer technology could help in the development of a universal language based on standard English in ways that go beyond merely improving the spelling. English has a number of other complications that could be eliminated altogether without sacrificing any intelligibility. The point is that, apart from its spelling, English is already sufficiently regular that the changes needed to make it perfectly regular are small; nearly all of them could be immediately comprehended by someone already fluent in standard English. The problem of not having a large enough population that can speak and understand a new language would thereby be circumvented. All we need to do is identify the irregularities that remain in standard English and devise changes so clever as to be perfectly regular and immediately comprehensible.

An example of a serious problem with standard English that may not be immediately obvious is that it has an unnecessarily large and redundant vocabulary. One interesting attempt to alleviate this particular problem was the development of Basic English by C. K. Ogden at Cambridge University in the early part of this century (Johnson 1944). The fundamental idea behind Basic English was to identify a subset of the English language small enough to be learned very quickly and yet large enough to comprise a fully functional language capable of communicating a broad range of ideas. What is most remarkable about Basic English is how small the language turned out to be: It has a vocabulary of only 850 words. It can be learned with fluency in only a few weeks, yet text written in Basic English is not readily distinguishable from standard English.

Basic English succeeded in producing a language that is easy to learn and at the same time easy for a fluent speaker of standard English to understand. But paradoxically, although it is easy for someone who is fluent in standard English to understand Basic, it is not as easy to speak and write the language, principally because such a person has instant and intuitive access to a far wider vocabulary than exists in Basic.

This suggests an important way in which computer technology might provide a critical service for the development of a universal language, and it is a service that the original inventors of Basic English in the early part of this century could not have imagined. One possible route to the proficient yet effortless use of written Basic English by a fluent speaker of standard English would be the use of computer technology to translate standard English to Basic. This may seem a somewhat odd suggestion because computerized language translation has been one of the more conspicuous and spectacular failures of the computer revolution. Very early in the computer era language translation was widely regarded as one of the important potential applications of the new technology. At that time it seemed that computers could be easily programmed to recognize the vocabulary and grammar of two or more languages, and they would then be able convert text from one to the other rapidly, accurately, and inexpensively.

Two of these three goals turned out to be as easy as expected: Computers today can translate languages rapidly and inexpensively, as the computer pioneers foresaw. Unfortunately, accuracy has proved to be a far more difficult problem. Languages are far too complex to be handled accurately by even the most sophisticated programs developed to date.

So why should it seem possible that computerized translating might be useful here? Simply because translating between one language (that is, between English and Basic English) might prove to be a much simpler problem than translating between two languages. All the rules and vocabulary for Basic English are a carefully defined and well-known subset of standard English. Will this type of translation prove to be simple enough for a computer program to handle? We can find out only by trying. If even this level of translation cannot be attained, we might at least learn something about the inherent difficulties of the translation problem. But if this translation should prove possible then the use of Basic English by both native speakers of English and by new users would be greatly simplified and enhanced, perhaps leading quickly to a very broad, if not universal use of the language.

Basic English, however, was an attempt to bridge the gap between non-English speaking and full fluency in standard English. No attempt was made to reform any of the major deficiencies of the English language, not even the spelling. And, as noted above, few changes would have as great an impact on the ease of learning the language as the phoneticization of spelling. But there are a number of other changes that are minor enough that they would barely be noticed by persons fluent in standard English, but which would make the language vastly easier to learn. The new language that would be created by these changes we might call "eng-lish" to distinguish it from standard English.

One obvious problem that could be easily eliminated

is irregular verb tenses. Verb tenses are themselves essential, but irregular forms of those tenses are nothing but useless impediments to learning the language. Fortunately, English verb tenses are already reasonably regular. The changes needed to regularize *all* of them are so slight that the regularized forms could generally be easily and immediately comprehensible to fluent speakers of standard English. The simplest way to regularize verb tenses would be to use the infinitive form of the verb without any changes except for the presence of different and uniform auxiliary or "helper" words that distinguish the tense. (This technique is very similar to the use of prepositions to reduce the number of inflections of words, a concept that has already greatly simplified the learning of standard English.) The obvious advantage of this technique is that a student needs to learn only about a half-dozen universal tense forms. Then as soon as the infinitive form of the verb is learned, its full conjugation is immediately known. The following is a sample of how it might work:

Infinitive: to see
Present: see
Past: did see
Past perfect: have see
Future: will see
Future perfect: will have see
Subjunctive: (who needs this?)

Nothing of any value is lost by deleting a few superfluous *n*'s, or even a few superfluous tenses. Some of these constructions may sound a little odd to speakers of standard English, but they are nonetheless instantly comprehensible. And the only penalty to be paid for this simplification is that it occasionally requires the use of an extra word, as in "did see" instead of "saw." But this is precisely analogous to the penalty paid for using an alphabetic script instead of a pictographic one like Chinese. It may take a little more time, effort, and space to write "antidisestablishmentarianism" than its equivalent Chinese character, but the added trouble is more than adequately compensated by the ease of learning the system.

Another of the important changes that could be made in the english language is the elimination of all inflections, the case-endings that indicate the relations of words in a sentence in many languages—subject, verb, object, and so on. English is already one of the least inflected languages in the entire Indo-European group. The inflections that remain are historical accidents that generally perform no important function and convey little or no information, but remain as useless relics of formerly necessary rules. The relic inflections in English are the subjective/objective case (in pronouns only), singular/plural, and the possessive case.

The distinction between subjective and objective case in nouns is of critical importance in many languages, including some of the root sources of the English language. In these languages the inflection is the only way to distinguish the function of words within sentences. But English uses word order and prepositions to perform the same function, and the system works well enough that the distinction between the subjective and objective cases has been dropped completely in English nouns; it survives only in relic form in pronouns, where it no longer conveys any useful information. This is obvious from the variety of jokes that exist on the usage of who/whom. The subjective/objective case distinction should be dropped completely in english pronouns.

Another useless inflection that survives in English is the singular/plural case. English has already simplified this inflection by dropping a third case, dual (distinct from singular and plural), that exists in many Indo-

European languages. To see how useless this remaining case distinction really is, consider extending the earlier system to have singular, dual, ternary, quaternary... and so on cases, that is, separate and distinct case endings for every possible number. In fact the amount of information conveyed by the plural case in nouns is so slight that it is not found in many languages, including Chinese. "One cat, two cat, many cat" is a perfectly comprehensible construction, and it eliminates not only an unneeded case distinction but also a variety of irregular case forms such as child/children and deer/deer (in which the inflection has already been dropped, without noticeable loss).

If the plural case in nouns conveys little information, in verbs the distinction conveys no information at all. It is perfectly useless, as Pinker noted:

> Take the English agreement suffix -s as in *He walks.* Agreement [between subject and verb] is an important process in many languages, but in modern English it is superfluous, a remnant of a richer system that flourished in Old English. If it were to disappear entirely, we would not miss it, any more than we miss the similar -est suffix in *Thou sayest.* (1994, 43)

Elimination of the useless singular/plural inflection in verbs would get rid of a slew of grammatical conundrums such as are described in Strunk and White (1979, 11), for example:

"The Republican Headquarters *is...*"

but

"The general's quarters *are...*"

Yet another odd example of a useless distinction in standard English is the difference between "between" and "among." This may be the sole surviving relic in standard English of the ancient distinction between the dual and the plural case, and it also serves no useful purpose: It is perfectly comprehensible to say both "between the two of us" and "between the five of us." Indeed, many fluent users of standard English have already adopted this convention.

The only remaining inflection in standard English is the possessive case. Eliminating it from english would get rid of a variety of minor difficulties that arise when a noun ends in *s* (as "the rhinoceros'(s) horn") or when the possessive and the plural case are combined. It might seem clumsy to have to say "the horn of the rhinoceros," but speakers of the French language already do this, and it does not seem to trouble them greatly.

English has also lost the gender distinctions between common nouns that have no obvious sex. This is a practice that is common in Indo-European languages, familiarly in French and German today. These distinctions convey no information but require a great deal of useless memorization: What possible point is there in having to remember whether a ball-point pen is male or female? English has profitably dropped such absurdities. But english could go further. It could drop the remaining gender distinction in pronouns. This would not only make the language more "politically correct," but would immediately eliminate a variety of conundrums such as "Everyone had his/her/their dinner." (All of these forms are sometimes used.) There is also no need to comment at any length on the elimination of other truly noisome gender distinctions such as aviator/aviatrix, alumnus/alumna, actor/actress, or waiter/waitress.

Distinctions of word forms for age brackets could be eliminated to get rid of a slew of irregular constructions (child, duckling, fawn, lamb, etc.). Young human, duck, deer, sheep, and so on would be a perfectly regular construction, already understandable to an English speaker, and far easier for someone else to learn. This is more a matter of simplifying the vocabulary than the grammar, however.

Yet another useless distinction that should be dropped from the english language is the difference between adjectives and adverbs. There is no difference in information content between "He played well" and "He played good." The english language should have a category of "modifiers," combining the two classes, and applying equally to nouns, verbs, or other modifiers as appropriate.

Finally, prepositions could be regularized. The preposition is a marvelous device that has enabled the English language to dispense with most of its inflections, and the use of prepositions is generally quite regular and clear. But there remains a small but regrettable number of irregularities. For example, we say, "I'll be there *on* Monday," but "...*in* January." It would not be so bad if this were just a case of completely synonymous and interchangeable words, but we never say "in Monday" or "on January." Similarly, we say "as large *as*" but "larger *than*." There is no need for different prepositions in these constructions. To really understand how useless these distinctions are, try explaining the reason for them to someone who is not fluent in English.

This discussion is only a beginning—I do not have all the answers for how best to create a regular, easily understood, and easily learned universal language based on standard English. I wonder whether we can get along without participles and gerunds, for example, or whether it might be possible to regularize these verb cases without sacrificing much intelligibility. When it comes to a choice between regularization of the language form and keeping it intelligible, my preference would be to favor the regularization, because intelligibility will be an issue only during a transition period to the new language, whereas questions of regularization could linger for rather long periods of time. Of course if we never get to the transition period because the language is never adopted, then the entire effort will be futile. We therefore cannot ignore intelligibility questions altogether. In the interest of intelligibility it might be necessary to declare all of standard English to be a legal segment or superset of english during a transition period, during which time the nonregular portions of the root language would be considered to be correct but archaic forms.

Compiling a complete list of the irregularities that should be eliminated from standard English will require a large amount of detailed research. Professional linguists and translators might be tapped for ideas. Also, English language teachers are probably familiar with many of the irregularities, having observed their students' difficulties; they could be canvassed for further ideas. Other profitable sources of information are students themselves; every nonnative speaker of English probably has a personal list of bêtes noires in the English language. A colleague of mine in France pointed out the "as-than" irregularity above, noting that French uses "que" for both constructions. With some diligent research along these lines we should be able to compile a reasonably complete list of irregularities in standard English.

And if it should prove possible to use computers to translate from standard English to Basic, then it should similarly be possible to use them to translate to english, for the same reasons. The english language could then become the universal "second language" even for speakers of standard English, and from there it would be a short step to universal usage.

One final consideration would be the necessity of having sufficient richness in the language to meet all of the various needs of users. For example, it would probably not be possible to write great poetry in Basic English. Any attempt would probably have more curiosity value than genuine literary merit, rather like the famous novel that was written without the letter *e*. However, an english language with all of the irregular spellings, verb tenses, and so on removed would immediately possess all of the richness of standard English, and should be adequate for poetry, literature, drama, and all of the artistic uses of language in addition to the more utilitarian functions.

Some have suggested that revisions to the English language would be rejected by lovers of literature because the revisions would make it difficult to read and appreciate the extant body of English literature. But the changes suggested here are small enough that most of this literature would remain as accessible as Shakespeare is today—readers would only have to familiarize themseleves with a few archaic words and grammatical forms.

A far more vexing problem is that most of the world's literature is already accessible to many of us only in translation. For myself, I would very much like to read Goethe, Proust, Cervantes, Dante, Euripides, Li Po, and other writers in their respective original languages. I find it distressing to think that this is not likely to ever happen. Even the scriptural texts of the major religions of the world—the Bible, the Koran, the Vedas, and so on—are inaccessible to many adherents in their original language.

Nothing can be done today about the plethora of languages that were used for literature in the past, but we should try to rectify the problem for future generations.

Since nearly the entire corpus of literature has yet to be written, the translation problem could be minimized for our descendants.

One might think that translations are all that is needed to allow readers to understand and appreciate the literature of an unfamiliar language. But anyone with experience in the area knows how inadequate, disappointing, and frustrating such translations can be. Edith Hamilton (1937, 12) writes: "I believe that the best a translator can hope for… is to convey some of his own enthusiasm, something of the impression the poem made upon him." A couple of examples will illustrate the difficulties that are faced by someone who has not studied classical Greek but would like to read Aeschylus. In the final line from *Prometheus Bound*, Prometheus, chained to a rock, with an eagle coming to tear out his viscera, cries out to the Gods of Earth and Sky; Grene (1956, 179) renders Prometheus's cry as:

You see me, how I suffer, how unjustly.

Whereas Hamilton (1937, 143) translates the same line as:

Behold me. I am wronged.

Both lines mean roughly the same thing. But for me their emotional content and impact are vastly different. Hamilton's line resonates far better with the rest of the drama, whose theme centers on heroic defiance and resistance to tyranny. Again, earlier in the play, there is an important passage that Grene (1956, 149) translates as:

…misfortune
wandering the same track lights now upon one
and now upon another.

Whereas Hamilton's lucid translation runs (1937, 107):

...Remember,
trouble may wander far and wide
but it is always near.

Grene's translation seems to imply that Prometheus's misfortunes are somehow happenstance, while Hamilton's lines convey a sense of menace, doom, and the smoldering enmity of Zeus. I vastly prefer Hamilton's translations, and I hope that they are faithful to the original text both in content and in visceral punch. But I have no way to know with any certainty without first mastering classical Greek. To further master *all* of the languages that have a literature that would interest me is a task that is beyond even the greatest linguistic genius. And these examples elucidate only a tiny fraction of the difficulties involved in trying to read and understand a work in translation. We should note that if learned and careful human translators cannot agree any better than this, then it is small wonder that computers cannot translate effectively.

Having a single universal language would not only eliminate the problems of translation for the future, it would enormously simplify the problems of translation of ancient literature. The gain in efficiency is similar to the way that parcel delivery services are made more efficient by moving all parcels through a single hub airport no matter what their origin or destination. The arithmetic is simple: If there are 3,000 languages, and it is necessary to translate among all of them, then 3,000 different translations would be needed for each text. In contrast, only a single translation would be required for each work to be brought into a single universal language.

Thus the need for a universal language has been clear for centuries, and all past attempts have met with little except varying degrees of failure. But modern communication technology (to say nothing of transportation technology) has made the need more urgent. And the same technology, applied to translation of a new universal language and coupled with the use of voice-actuated word processing to promote phonetic spelling, could promote the creation and use of such a language today, if we were to make the necessary effort.

7

Decimal Delenda Est

In exploring the practical effects of the computer revolution we have to accept the fact that some of its most far-reaching consequences are not going to be easy to foresee at the present time. Many of the subtle features of the computer revolution, features that may seem inconsequential today, could have drastic long-term effects. For example, computers do not use decimal arithmetic. They do all their arithmetic using something called binary notation, which we will examine in more detail shortly.

This chapter will explore some of the possible consequences of the singular fact that, for all intents and purposes, decimal arithmetic is no longer in use in our civilization. It is not much of an exaggeration to say that all arithmetic today is done in binary (by computers). Of course these same computers normally translate their final results into decimal for the benefit of their human users, and most users are therefore blithely unaware that any conversion has ever taken place. It might seem that this is the end of the matter, that there should be no further consequences of the fact that computers are using a better arithmetic system than we are. But it is likely that such a dual system of arithmetic can prevail through the indefinite future? I rather doubt it.

In point of fact the argument of the previous chapter, that we should modify the English language to meet the needs of the computer age, applies even more strongly to decimal arithmetic. We should not continue to use an obsolete relic from the Middle Ages when far superior arithmetic systems are available. These systems are vastly simpler both to learn and to use than conventional decimal arithmetic, and, as an added bonus, they are much more compatible with computer usage. Indeed we might argue that by not using decimal arithmetic our computers are exhibiting more intelligence than we are.

The original introduction of decimal or Arabic arithmetic to Europe by Leonardo of Pisa (a.k.a. Fibonacci) in the early part of the thirteenth century was undoubtedly one of the pivotal events in European history. The decimal arithmetic system was so far superior to the Roman system that was then in common use that for a long time no one asked whether an even better system might exist.

The critical idea that sets Arabic numeration off from earlier systems such as Roman numerals is the idea of place-value notation. With place-value notation, the first digit (from the right) represents the number of units, the second digit gives the number of 10s, the third digit gives the number of 100s (10^2), the fourth digit gives the number of 1,000s (10^3), and so on. Thus the decimal number 2,843 is interpreted as:

$$(2 \times 10^3) + (8 \times 10^2) + (4 \times 10^1) + (3 \times 10^0)$$

(Any number raised to the 0th power is 1, of course.)

This place-value system is ingenious, powerful, and of fundamental importance, but the value of 10 is not. Any positive integer can be used. To understand what this means, take a simple example: The binary system uses 2 as the base or radix, so 1,011 in binary is interpreted as:

$$(1 \times 2^3) + (0 \times 2^2) + (1 \times 2^1) + (1 \times 2^0)$$

In fact there is no need to limit the options to positive integers. Any number except 0 could be used for the radix, even a negative or an irrational or a complex number. The use of a complex radix leads to a particularly lovely branch of arithmetic in which ordered pairs of real numbers are no longer needed to express complex numbers, but that discussion would take us a bit far afield of the main point here. The details can be found in Knuth (1981).

But if any number could be used for a radix, then the obvious question arises of which number would provide the "best" radix. What are the advantages and disadvantages to the various options?

Mathematicians have long debated this last question. Pascal argued as early as 1665 that 10 is a poor choice (Glaser 1981, 173). Some have advocated the use of 4, 8, 12, 16, or even 60 (!). There is even a society (the Duodecimal Society, presently based in Garden City, New York) that was set up to promote the use of base 12. Not surprisingly, these efforts have borne little fruit, because until the invention of electronic computers the advantages of one radix over another were relatively minor, and there are obvious costs involved in any change. But of all the possible choices, 10 turns out to be one of the worst. Base 10 does not possess even minor advantages. It has almost no redeeming virtue other than long habit, a habit that is so strong that advocates of change have often despaired of ever making progress.

Why was base 10 chosen in the first place if it is such a bad choice? Although no records survive of the original invention of decimal arithmetic, the choice of 10 clearly was based on nothing more important than the number of fingers we use to count on. The choice was certainly not based on any rational analysis of the advantages of the different choices of base; even the possibility of other choices was not realized for several centuries after the adoption of base 10.

With the invention of the electronic computer the situation suddenly changed. Today, as we noted above, the use of decimal arithmetic has essentially vanished; virtually all arithmetic is done in binary by computers, and the results are converted to decimal when the calculation is finished. This system is not terribly efficient, but it works tolerably well. Why should we want to change it? To answer this question we need to survey the various advantages and disadvantages of different choices of radix.

The most commonly cited advantage relates to the number of factors that the radix possesses, since numbers (especially fractions) related to a factor of the base tend to have simple representations. Thus it has often been suggested that base 12, with its factors of 2, 3, 4, and 6, would be an improvement over decimal arithmetic. In base 12 arithmetic the "decimal" (duodecimal?) representation of 1/3 is 0.4, instead of 0.33333... in decimal. Although this is the principal reason cited for the choice of base 12, in fact this advantage is relatively slight, since there is no choice of base that will make all fractions simple, and all such choices will simplify only a very small percentage of fractions. Six would actually be a far better choice than 12. Base 6 possesses all of the advantages of base 12 for simplifying fractions, and has a far simpler arithmetic as well. There is in fact no reason at all to choose 12, and, as we shall see, there is an even better choice than 6.

The more important advantages and disadvantages of the various choices of base stem from the number and variety of digits required to express a number in each base. Each base system requires as many distinct digits as the base itself; decimal requires ten digits, octal (base 8) requires only eight. Larger bases with more distinct digits have an advantage in that they generally need fewer digits to express any given numerical value. But this advantage is offset by an increase in the number of basic arithmetic operations that must be memorized to manipulate those numbers. This disadvantage is serious, because the growth in the number of arithmetic operations is fairly fast (it grows as the square of the base), while the decrease in the number of digits for a given numerical quantity is quite slow (it decreases as the logarithm of the base). For example, if you triple the size of the base you increase the size of the multiplication tables by a factor of nine, but you decrease the number of digits in a given number by only about 50 percent.

This argument suggests that we should use the smallest practical base, which is 2. And base 2 has important advantages for computers, as John von Neumann first pointed out, because binary notation requires only two distinct digits (0 and 1) that can be represented in the electrical circuits simply by on-off states. In point of fact it is possible to design nonbinary electronic circuits in which different numerals are represented by discrete voltages, for example 0–9 volts for decimal arithmetic. Although such circuits have been constructed in the laboratory, they have not found wide use. The added complexity (and cost) of the equipment required to generate and discriminate among the different voltage levels has generally outweighed any slight advantage gained from using nonbinary arithmetic.

Thus for the first time in modern history a nondecimal (binary) arithmetic is essentially in universal use. But even among computer experts binary arithmetic is generally left to the machines. Even though binary arithmetic is simple to the point of being trivial (the entire multiplication table has only three entries, and *all* of them involve the trivial operations of multiplication by 0 or 1), the binary numbers themselves are so large and cumbersome that manipulation of them is both difficult and error-prone. The following table illustrates the problem:

Decimal	Binary
1	1
2	10
3	11
4	100
6	110
10	1010
25	11001
100	1100100
697	1010111001

The binary representation of 697 contains more than three times as many digits as the decimal representation, and the lack of variation among the digits (only 0s and 1s are used) makes binary numbers especially difficult to memorize or manipulate. Most of us, for example, are able to glance at the number 697 and hold it in short-term memory long enough to use and manipulate it. But the binary equivalent, 1010111001, requires significantly more effort. To take another example, the house I grew up in had the unremarkable street address of 8400. Imagine the difficulties for the postal service if this number had to be expressed in binary: 10000011010000. Thus, except for very specialized applications, binary arithmetic is seldom used even by computer programmers today.

As we noted above, the most common way to avoid the use of binary numbers when working with computers is to let the computer convert its input and output (its interaction with the user) to decimal while the computer operates internally in binary. The computer does all the work of conversion; the user is often blithely unaware that the conversions have even taken place. This system works reasonably well in most cases and is by far the most common means of interacting with computers. But some serious problems are involved in the conversion of decimal numbers to binary, as most programmers eventually discover. Binary notation is simply not an efficient code for representing decimal digits. Some of the problems can be seen in the difficulty of converting fractions discussed earlier. Just as 1/3 has an exact (finite) representation in base 6 or 12 but not in base 10, a fraction such as 1/10 has an exact representation in base 10 (0.1) but not in base 2, where it looks something like 0.0001100110011...(the pattern 0011 repeats indefinitely).

How can we avoid this type of conversion problem and still have the computer use binary arithmetic internally? The usual method is to use a base that is an exact power of 2. Conversions from binary to base 4, 8, 16, 32, and so on, are quick and efficient, and avoid all the problems relating to fraction conversion. We can illustrate how easy and convenient it is to convert to such a base by converting a number from binary to octal. All we need to do is to divide the binary number from the right into groups of three digits each, and read each group as a binary coded octal digit. Thus:

2	1	5	7	6	3	(octal)
010	001	101	111	110	011	(binary)

So 215763 in octal equals 10001101111110011 in binary. Similarly, to convert to hexadecimal (base 16) we would divide the binary digits in groups of four rather than three. The magic number for the size of the groups is simply the exponent on the base conversion. In other words, $8 = 2^3$, $16 = 2^4$. Exactly why this technique works, and why it requires the radices to be related by an exact power will be left as an exercise for the interested reader.

The important point is that for simple and efficient conversion to binary arithmetic we must choose a base that is an exact power of 2, and 8 and 16 have been extensively employed by computer programmers for this purpose. But base 4 appears to be a far better choice than either 8 or 16, and, oddly, it is a choice that is almost never used. Eight and 16 appear to have been chosen for no better reason than that they are closer to 10 than any other power of 2. In other words, through these choices the mistake of the thirteenth century is being propagated into the twenty-first century.

The disadvantage of hexadecimal is simply that its arithmetic is much more complicated than decimal. Its multiplication table, for example, is more than two and a half times larger than the familiar decimal multiplication table. The disadvantage of octal is more subtle. In order to convert binary to octal the binary digits ("bits") must be grouped in sets of three. There is nothing inherently wrong with using groups of three bits, but very few computer systems or digital communication systems are organized into binary words of three bits or any multiple of three bits. Most computers use word lengths that are some multiple of eight bits, either eight or sixteen or thirty-two bits, and digital communication is often carried out with eight-bit word lengths. In fact, the eight-bit binary word is in such common use that it has been given a name of its own, the "byte." Octal arithmetic is simply not well suited for use with binary numbers that are organized into bytes.

Quartal arithmetic would avoid the disadvantages of both octal and hexadecimal. Conversion of binary to quartal involves grouping the binary bits in pairs, and so the "dangling" bits involved in the conversion of bytes to octal digits are completely avoided. The lone disadvantage to the use of quartal arithmetic is that its numbers tend to be slightly longer and clumsier to use than those in higher base systems. However, the number of additional digits that must be dealt with is less than one might think. Quartal numbers use about twice the number of digits as the corresponding hexadecimal number, but less than twice the number of digits of the decimal number. To illustrate the point, let me expand the earlier table. (Hexadecimal notation uses the letters A–F to represent the digits with values of 10–15.)

Decimal	Binary	Quartal	Octal	Hexadecimal
1	1	1	1	1
2	10	2	2	2
3	11	3	3	3
4	100	10	4	4
6	110	12	6	6
10	1010	22	12	A
25	11001	12	131	19
100	1100100	1210	144	64
697	1010111001	22321	1271	2B9

The notational simplification of quartal over binary is significant, and the presence of four distinct digits rather than just two aids in memorization and manipulation. The advantage resulting from the smaller number of digits in either octal or hexadecimal seems to me to be slight.

The quartal multiplication tables are 5.5 times smaller than the corresponding decimal tables. But the real computational advantage of quartal over decimal is even greater than this because 70 percent of the quartal tables involve trivial operations with 0 and 1. If we count only those addition and multiplication operations that do not involve 0 and 1, we find that the quartal tables are fully twelve times smaller than the corresponding decimal tables. In fact, the entire nontrivial portion of the quartal multiplication table contains only three entries: 2 x 2 = 10; 2 x 3 = 12; and 3 x 3 = 21; memorize these three facts, and you have mastered multiplication in quartal arithmetic. This advantage was noted by Thiele in 1889 (see Glaser 1981, 173). And the only price to be paid for this enormous computational simplification is that the quartal numbers tend to have about 50 percent more digits than the corresponding decimal numbers.

Another advantageous feature of quartal arithmetic, noted by Stein in 1826 and Hankel in 1874, is that it requires fewer fundamental concepts to express any given number up to 1 million (again, see Glaser 1981, 173).

One more minor but interesting feature of quartal arithmetic involves the familiar rule in decimal arithmetic used to test for divisibility by 3: If the sum of the digits of a number is divisible by 3, then the number itself is divisible by 3. The rule happens to work in quartal arithmetic as well, but not in octal. It does work in base 16, and, curiously, it works for both 5 and 15 as well as 3 in base 16. In general, in base N, it works for $N - 1$ and any factors of $N - 1$. Thus it works for 7 and not 3 in base 8.

For diehard users of hexadecimal arithmetic, conversions from quartal to hexadecimal are just as easy as from binary to quartal. Pairs of quartal digits convert to single hexadecimal digits the way pairs of binary digits convert to quartal digits. Users of octal will have a slightly more difficult time: Triplets of quartal digits convert to pairs of octal digits. The easiest way to understand this is to convert both the quartal and octal numbers to binary. Even this conversion is easier than from octal to hexadecimal, where triplets of hexadecimal digits convert to quadruplets of octal digits. In the conversions between binary, quartal, octal, and hexadecimal, octal is clearly the odd man out.

Prior to the 1940s practically nothing came of the many suggestions to change from decimal arithmetic. Yet today binary is in universal use for essentially all arithmetic operations. In the span of two generations the use of decimal arithmetic has essentially vanished, and, amusingly, almost no one is conscious of this. Only we humans persist by long habit in our ancient mistake, but we are no longer performing any significant fraction of the arithmetic done in our civilization.

At the same time octal and hexadecimal arithmetic are in fairly common use among computer programmers, and that usage will increase as computers become more and more common. If we are to avoid repeating the old mistake then we might do well to at least try using quartal arithmetic for those applications where octal or hexadecimal are used today. The advantages of quartal are so great that quartal would probably become the de facto standard in the computer industry if it were widely available. Software manufacturers might find that the minor changes required to offer quartal I/O as an option would provide an interesting way to develop a competitive advantage.

The adoption of quartal by the computer industry could prove to be a useful first step toward a more general use of quartal arithmetic. If quartal were adopted by computer users, we might soon develop a generation of young students who are equally comfortable with both quartal and decimal. They might then prefer quartal to decimal just as people familiar with both decimal and Roman numbers usually employ decimal, and for the same reason—ease and simplicity of calculation. Then in a mathematical inversion of Gresham's law, we might find that the "good" arithmetic would drive out the "bad." Decimal arithmetic could then be reserved for ceremonial purposes such as carving on buildings and monuments.

The difficulties and costs involved in making such a switch are, of course, formidable. They are worse than the difficulties involved in converting to the metric system of measurements, and the United States has made little progress to date in adopting metric measurements even though we have officially been on the metric system since the Jefferson administration. As one wit put it, progress toward metric is inching along. However, one of the advantages to be gained from a general conversion to quartal arithmetic is that it would make the failure to convert to metric irrelevant. The metric system itself would become obsolete. In quartal arithmetic it is easier to convert from feet to inches (multiply by 30) than it is to convert from meters to centimeters (multiply by 1210). Of course, it would be simple to modify the metric system so that it would use quartal arithmetic effectively, but an entire new system of size-related prefixes would have to be devised.

The world is unlikely to soon adopt quartal arithmetic. But people who are currently using octal or hexadecimal arithmetic should consider switching that usage to quartal. If there are to be two systems of arithmetic in use, then the second one ought to be quartal rather than octal or hexadecimal.

8

The Computer Revolution in the Arts

The arts represent another broad area, like language, that might appear to be immune to technological revolutions. Many artists are notoriously technophobic. But the arts are no more immune to technological changes than any other aspect of civilization. Technology has always been the handmaiden of art. Consider, as an obvious example, how many ways electronic technology has already altered musical expression through inventions as diverse as electronic amplifiers, electric guitars, and compact disk recordings. Nor are painters restricted today to the technologies that were used to create the extraordinary works of art in the caves at Lascaux.

Artists have always pursued and experimented with new media and modes of expression that have been made possible by advancing technologies. And technology is about to create new possibilities on a scale that is utterly without precedent. The new electronic techniques could produce the most far-reaching changes ever encountered in virtually every form of art. As with education, art will not be locked into technologies that were inherited from earlier centuries.

And although the human race is likely to produce new Mozarts and Leonardos who will be able to fully exploit the opportunities presented by the new technologies, the necessary genius will not necessarily appear immediately. A great deal of experimentation will have to be done before the capabilities of these new technologies are fully understood. And, as has always been the case, much of the experimentation will be worthless; nearly all of it will be simply junk. As Aristotle famously remarked, even Homer nods occasionally. And lesser artists nod off more frequently.

Indeed, one of the principal reasons for the enduring appeal of classical art and music is that it has already been though a severe selection process—much of the junk was weeded out long ago. The classical works that have endured to the present represent only an infinitesimal fraction of what was originally created. Only the finest art and music has survived the long winnowing process. In the exploitation of novel electronic media our primary interest will once again be in the very finest work. We will have to ignore the rest, which, unfortunately, will comprise the vast bulk of it.

This discussion can explore only a few of the new possibilities introduced by burgeoning technologies. The catalog is not intended to be complete, but rather is a mere sample of what will become available. Many of the relevant technologies have not yet been fully developed; many of them quite likely have not even been conceived.

We could begin by exploring visual expression, the creation of two-dimensional images. Classical technologies for visual expression include such things as painting, drawing, and etching. One obvious new medium for visual expression is the computer display screen. Unfortunately, present display screen technology is extremely limited as a medium for artistic expression, even though this primitive technology has already found wide use in graphic design. The critical problem is that present CRT (TV-tube) displays have some very serious limitations—they are restricted in size, in resolution (pixel size), and in color rendition and accuracy. They are also bulky, fairly expensive, and consume a relatively large amount of power. Fortunately, technologies are being developed that will eliminate all of these problems. The new flat-screen displays use a variety of new technologies, including such things as liquid-crystal displays, active-matrix technology, and electroluminescence.

Which technologies will prevail is uncertain. Nevertheless, we might reasonably expect that new technologies will eliminate all of the problems of conventional CRT displays. New computer displays that combine large size with fine resolution and clear, accurate, and reproducible color rendition should provide a powerful new medium for artistic expression. Such a medium should allow artistic possibilities that were impossible with conventional images on paper, including images that vary with time, or even variations that are sensitive to the presence and configuration of viewers.

Other new technologies are being developed that could blur the difference between visual (two-dimensional) representation and sculpture (three-dimensional representation). The most obvious of these technologies is holography, with its capacity to produce 3-D images. Still other techniques are also being developed that use laser scanning to produce genuine three-dimensional images, not just 3-D illusions. Conceptually these techniques can be thought of as three-dimensional CRT displays utilizing phosphor "volumes" rather than phosphor "screens."

But the technology that may offer the most dramatic possibilities for both 2-D and 3-D artistic expression is the set of techniques that are broadly referred to as "virtual reality." Virtual reality techniques involve the computerized simulation of external objects. With current virtual reality technology special goggles are used to project computerized images in front of each eye and thereby simulate the 3-D visual experience of objects that do not actually exist. The projection technology is coupled with sensors that allow the computer to monitor the user's motions. As users move or turn their heads the computer changes the projected images to correspond to these motions.

In addition to visual virtual reality effects, aural simulation can be added using earphones instead of goggles. And some experimentation has been done with tactile simulations employing mechanized gloves worn on the hands or even entire mechanized body suits. Tactile simulation is a far more difficult problem than visual or aural, and the technology for it is far less advanced and less effective. Nevertheless all three technologies are being developed, and there is no reason not to expect the usual cascade of technical improvements in the near future.

Virtual reality technologies have been developed largely for games and simulations, but they could easily be exploited for artistic purposes. Such techniques allow the possibility of sculpture that would be freed from the mundane constraints of the physical world: The sculptor could create artworks that fly or balance on a pinpoint. The sculpture could be kinetic in ways that would have astonished Alexander Calder. And stunning possibilities exist for combining virtual reality techniques with art forms such as dance or grand opera. The possibilities are endless, especially if tactile and other forms of sensory simulation become practical.

For real sculpture involving the production of genuine three-dimensional objects rather than just images and simulations, computerized manufacturing techniques are being developed that can produce arbitrary three-dimensional shapes using laser-controlled sintering techniques. These techniques are being developed primarily for precision-automated manufacturing, but they could also be applied to artistic endeavors. These computerized manufacturing techniques offer the promise of creating physical objects with a level of precision and control that is impossible with manual techniques.

In music the impact of computer technology could be just as dramatic as in the visual arts. The dichotomy between composition and performance that has been characteristic of music for generations may disappear altogether. It will exist no more in music than it does in painting. The concept of a "performance" by human musicians will become obsolete as computerized music far outstrips the capabilities of the performers. Those performers will be in a position somewhat analogous to horse-and-buggy drivers in the era of the first horseless carriages. And like those horseless carriages, computerized music today is the subject of some contempt and derision, but as it continues to improve, human performers will be unable to keep up with the pace of the progress.

It may be a good thing that we have high-quality recordings of some of the greatest violinists, pianists, trumpet players, and other instrumentalists now alive, for future musicians may have little incentive to devote the years of effort that are required to master these instruments. They are as likely to be found in musical performance in the future as the dulcimer and sackbut are today. Of course some composers and performers will use the old instruments and techniques, just as some people enjoy calligraphy in an age of typewriters and laser printers.

A foretaste of the capabilities of digitized music was found in a technology that was popular a century or so ago: The player piano was actually one of the earliest digital technologies used for musical performances.

And in general, player pianos were designed and used to reproduce (with varying effectiveness) conventional musical pieces written for the piano. But a small genre of compositions was written exclusively for the player piano. Such works employed "impossible" combinations such as chords that span a range that is too broad for the human hand or even chords that involve more than ten keys at the same time. These compositions transcended the limits placed on piano performances by the physical limitations of even gifted pianists.

Two fundamental limitations of musical performances can be eliminated with computer technologies. The first is caused by the limitations of the musical instruments, and the second by limitations of the human performer. Of course we have some marvelous, almost magical conventional instruments, such as the violins made by Stradivarius, and equally phenomenal and talented performers who have created marvelous musical performances. But the extraordinary quality of such musicians and instruments should not blind us to their limitations.

The limitations of individual instruments and performers have been circumvented in the past in many ways besides the specialized use of player pianos. The most obvious is the use of multiple performers, duets, quartets, and, perhaps most impressively, symphony orchestras. But computer technologies will allow the creation of any sound pattern that can be conceived, patterns that have far greater variety than can be produced by any combination of human performers and instruments.

Technologies have already been developed for computerized musical composition and performance. The MIDI standard (Musical Instrument Digital Interface) contains specifications for both a new (digital) standard for musical notation and composition and also standards for networking of different computerized music generators, keyboards, synthesizers, and so on. The MIDI standard allows the composer to specify tone, pitch, timbre, duration, volume, and other qualities of a musical sound with greater precision and flexibility than is allowed by conventional notation. It should be possible, for example, to specify a different volume level for each note within a piano chord. With MIDI technology the composer can edit the resulting composition as if it were text on a word processor, and can sample or "perform" the resulting work instantly, even work that would normally require an entire orchestra.

Of course nearly all the sound patterns that are mathematically possible are worthless, essentially indistinguishable from noise. Indeed, they are noise. But some fraction of them are something else. Some sound patterns represent musical possibilities unimagined by conventional composers and performers. They will become the standard material for musical work when their quality, their aesthetic appeal, and their basic interest for the audience exceeds what is possible with conventional instruments and performers. Some hint of what is possible was first heard with the synthesizer craze of the '60s and '70s. But the computer technology available at that time was primitive, so weak that the synthesizers were in some ways more limited than conventional instruments. Modern computer technology will allow much more flexibility in the specification and creation of sounds, tones, timbres, and other audio effects.

Computer technology may affect the production of even vocal music (singing), through the use of voice synthesis. The current state of computerized speech synthesis is so primitive that we can hardly imagine its use for the creation of music. But it will only improve in the future. At some point it will become better than what can be produced by the human voice. It will then become a separate art form.

Indeed, many musical forms are already enhanced by digital recording technology to the point that they are significantly better than what can be produced directly by human performers. Recordings of musical productions can be edited using computer technology to correct single notes or replace even longer passages. The resulting production can attain a level of perfection that is beyond the capacity of human performers. I have heard one performer complain that such recordings have raised the standard that audiences expect of musicians. Eventually audience expectation will outstrip what even the finest and most talented performers are capable of achieving.

In addition to replacing performers and instruments, computer technology could have a transforming effect on the process of composition. Classical composers create musical scores on paper, of course, but they generally do not think in terms of notes any more than authors think in terms of sequences of letters. More generally, they conceive of sequences of sounds, often with the help of instruments such as pianos. They then laboriously transfer those sounds to a written score. Computer technology exists today that can eliminate much of the tedium of creating the physical score. If the composer is in the habit of working at a piano, for example, then computer technology can directly record the keystrokes. Indeed, such capabilities are part of the MIDI standards. Computers can also analyze and record sounds directly, so that a composer who works vocally could have notes transcribed automatically as they are sung or just hummed. This would be similar to voice recognition technologies for literary composition, discussed below, but computer identification of tones for music recognition is a far simpler problem than identification of phonemes for voice recognition.

Other composition functions have algorithms that could be handled by computers. Changing the key of a score is an obvious example. Also, compositions such as fugues and canons that require prescribed and ordered patterns could be computer-assisted. And computers can check for obvious problems such as parallel fifths. In these functions computers are performing tasks similar to spelling-checking technology for language. And as with spelling checkers, the algorithms will not be perfect, but they will be sufficiently accurate to be useful.

Dance is another art form that would appear to be completely immune to technological advance. In the classic philosophical conundrum, how can you separate the dancer from the dance? The contribution of computer technology to dance is likely to be restricted to the process of choreography, where computer display technology holds promise for novel methods of conceiving of dances as well as new ways of teaching the resulting choreographed work. The choreographer could create computerized moving images that show what he or she expects of the dancers. The resulting images could be used to instruct dancers across the ages, just as a score by Chopin is used to direct the play of pianists today.

Indeed, the resulting computer-aided dance-imaging technology could develop into an art form of its own, perhaps almost inadvertently. It is not uncommon for things that were never originally intended as completed works of art to become museum-quality objects, including studies and preliminary sketches done by famous artists. Computerized aids for artists could have a similar fate.

Finally, dance could be affected through the use of virtual reality technology to produce an art form that has never existed before, much as the invention of motion picture technology created a novel art form.

The artwork of motion pictures was originally based largely on techniques that had been used for centuries on stage, but it moved on to include effects that are not possible on a stage, such as capsizing an ocean liner or portraying realistic dinosaurs. Other combinations of visual and aural techniques are already being developed under the rubric of "multimedia" work. These efforts have barely begun to explore the combinations of sensory effects that are possible with computerized technology.

Computerized special effects have played a significant role in motion pictures from *Star Wars* to *Jurassic Park*. And the next logical step from synthesized dinosaurs is computer-synthesized human actors. A word has recently been coined to describe such computerized actors: synthespians. As with many computer simulation techniques, the technology for creating synthespians today is primitive, unrealistic, and unconvincing, but it will only get better as the technology is developed. Eventually it will eliminate problems related to temperamental actors and their attendant high salaries. I understand that already a movie is being planned that will star George Burns, recently deceased, using computer simulation technology. In a sense, synthespian technology is a natural extension of the animation technology pioneered by Disney in classics such as *Snow White* and *Peter Pan*. And the Disney production of *Peter Pan* in particular exhibits many of the ways in which synthetic actors can exceed the capabilities of ordinary actors.

The effects of computer technology on literature could be as extensive as on any other art form. The most obvious changes will be brought about with the development of voice-recognition technology. The advantage of this technology is that we can speak words at a rate that is several times faster than we can write or type them. After all, we have had several million years to adapt to the use of voice technology, compared to only a few thousand years to adapt to writing, and only a hundred years or so to adapt to the use of typewriters.

The effect of voice-recognition technology will be to increase the productivity of writers by factors of two to five. Suppose this technology had been available to writers in earlier eras. Would five times as much literature have been produced? Or would the extra time have been used to further refine some of the works that were created? Some may bristle at the thought that masterpieces by writers such as Austen or Shelley could be improved. But perfection is no more attainable in the arts than it is in mathematics, and we saw in chapter 4 that even infinite amounts of time and effort may not be adequate to attain perfection in mathematics. As the celebrated mathematician John von Neumann put it: "Truth is much too complicated to allow anything but approximations" (quoted in Schroeder 1991, 367). Similarly, perfection in artistic endeavor can only be approximated. Further, many artists have had to work under time pressures that may have precluded the opportunity to polish their work even to the standards of which they are usually capable. Shakespeare in particular may have been under severe time constraints since he had to stage his dramas and act in them in addition to writing them. And as noted above, even Homer nods.

Of course some artists revise their work more frequently and assiduously than others. Some may not do it at all, as is suggested in the memorable scene from the movie *Amadeus* in which Salieri is stunned to discover that Mozart's handwritten scores contain no erasures. But most composers and writers revise their work extensively. A fascinating description of revisions to familiar poems can be found in the *Norton Anthology of English Literature* (1979, 2513–33). In Blake's familiar poem "The Tyger," we find the final line revised from

Dare form thy fearful symmetry

to

Dare frame thy fearful symmetry

And Shelley is seen revising from a first draft

Ah time, oh night, oh day
Ni nal ni na, na ni

to

O World o life o time
On whose last steps I climb

This type of revision is, of course, far easier to perform with word processing technology. As a drawback, the use of word processors may make it less likely that authors' and composers' earlier drafts will be preserved in the future. This fascinating chapter from the *Norton Anthology* may prove to be much more spare for literature that is created with word processors in the future.

If we assume that the final work by artists of Shakespeare's caliber represents a level of artistic perfection that even he could not have improved on, the possibility remains that he might have created additional plays or poems, if only he had had more time. In other words, many great masterpieces may have been lost forever because of the technological limitations that he and many others worked under. Of course we have no way to know what artists in the past might have accomplished had more modern technology been available to them. But it seems fairly safe to assume that these technologies will at least make life easier for artists in the future.

Word-processing technology has other features that further enhance the productivity of the writer. Spelling checkers and grammar checkers are already available, and in writing this very paragraph I have the use of an on-line dictionary and a thesaurus. I can also dial up the Internet to check on facts and references. Of course not all the time that computer technology will save for writers will be used by them to create more or better compositions, but if some of it is so used by some of them, some of the time, then it could have a large impact.

Another aspect of the literary art that might be impacted by computer technology involves the consumption of literature rather than its production, or, in a word, reading. How will the reading process be affected by computer technology? The idea that printed books will be replaced by computer display technology seems at first to be as ludicrous as the idea that computer displays will be used for artistic purposes, and for many of the same reasons. Computer displays are bulky, they have limited resolution, and they are somewhat fatiguing to look at for hours on end, which is required for serious reading.

But this will change as computer display technology improves. Computer displays will eventually match the size, resolution, and appearance of a printed page. Suppose further that the hardware needed to store and display the quantity of information now found in several books can be contained in a package the size and weight of an ordinary paperback. Such technology could revolutionize the process of reading. For example, when readers encounter an unfamiliar word in the text of a conventional book they are presented with a set of unpleasant choices: They can either ignore the word and hope that it is not too important to the text, thereby decreasing their comprehension level, or they can consult a conventional dictionary, improving comprehension at a serious cost in time and energy, not to mention interruption of the continuity of the reading process.

But suppose that the computerized "book" has an on-line dictionary, perhaps activated by a touch-sensitive screen. Touching the unfamiliar word could open a window that contains the dictionary definition. The loss of time and energy as well as the break in continuity would be considerably reduced, with no loss in comprehension. The same technique could be expanded to include explanatory notes as well as definitions, notes of the sort that most readers today require in order to make sense of numerous passages in Shakespeare, for example.

A computerized book could have a number of other advantages as well. It could be completely cross-referenced. Suppose, for example, you are reading a novel by Tolstoy or Dostoyevsky, and you encounter the usual problems with keeping track of a large cast of characters with unfamiliar names. With a computerized book you could have instant access to the previous reference to any character, for example, or to the first reference, or to notes or a summary of the principal characters. Your comprehension and enjoyment, not to mention your ability to tackle difficult and unfamiliar texts, could be vastly enhanced by such technologies.

Computer technology also might be used to develop completely original and different forms of literature. For example, conventional reading is a linear process, starting at the beginning of a document and generally proceeding straight through to the end. However, the techniques that are broadly referred to as "hypertext" allow one to read parts of a document in whatever order the user cares to define. It allows the ordering of the text to follow the reader's cares and interests rather than the author's. These techniques could allow the entire process of both reading and writing to evolve into something totally different from the way literature has been thought of since the days of clay tablets.

Any number of other art forms will be altered by the computer revolution, sports and athletics, for example. It may seem odd to consider sports an art form, but sporting activities range from things that have considerable aesthetic overlap with dance, such as figure skating, gymnastics, and springboard diving, to things such as football at the other end of the spectrum. And computer technology could have a strong impact on all these activities, primarily through computerized techniques for designing and choreographing sporting activities, similar to the effect on conventional dance. Some training techniques rely on computerized analysis of actual performance. And the synthespian technologies described above have obvious application to sports such as football. Synthespian football players would have a degree of perfection that is unattainable by ordinary athletes. (This would work only with televised football games or with virtual reality techniques.) This idea could be carried to a ridiculous extreme by producing synthespian track and field stars who could break a three-minute mile or a three-minute marathon, for that matter. These sports could suffer the fate of chess, in which ordinary humans are unable to match the performance of machines.

The possibility of computerized art forms holds other advantages beyond simply new modes of expression and ease of editing and revising. With digital techniques all the problems related to preservation and display of artworks would largely disappear. The candle smoke that obscured the Sistine Chapel frescoes for centuries would not be a problem in a digital world. Instead, it would be possible to guarantee that future generations could always view a work of art exactly as it appeared to its creator. Imagine seeing the sculptures of Phidias and Praxiteles today as they originally appeared, complete with their original paint. And other works by artists as prominent as Da Vinci have simply vanished, or exist only in copies made by other artists. Similarly, scholars today are unsure of the exact text of some of Shakespeare's work. Digital art and literature, although it would not be perfectly indestructible, would have a level of permanence that is unavailable with conventional media.

It might seem that computerized records of any kind are particularly vulnerable to destruction. Horror stories abound in which system and disk crashes have destroyed vital records or erased critical back-up tapes. Yet today's nonmagnetic storage techniques, especially read-only compact disk recordings, are nearly invulnerable to such catastrophes. And error-correction codes can be used that allow perfect reproduction of records that have been damaged or partly destroyed. Finally, the ability to easily make large numbers of exact copies of computer records is perhaps the most foolproof technique for rendering digital records invulnerable. If the records are distributed widely enough, the possibility of a single calamity destroying all of the copies is reduced exponentially.

It would also be possible to modify digital artwork for particular displays or milieus or for other aesthetic reasons without ever altering the original work. I am not suggesting plagiarism, but rather possibilities that are more along the lines of an opera composer borrowing a story line from literature, as in Verdi's operas based on Shakespeare's *Othello* and *Macbeth*. The original artwork, preserved on well-dispersed exact copies, would be essentially indestructible, but it still could be used in ways that are impossible with conventional media, and even in ways that were unimagined by the original artist.

All the technologies that I have discussed exist today, albeit sometimes in rudimentary form. I have not strayed into highly speculative areas, such as virtual reality based on direct nerve stimulation. Such technologies might drive the arts in directions completely unimaginable today. Thus this summary of the possible effects of computer technology on the arts is only a beginning. Future generations may feel about the accomplishments and the technologies of today much as we do about the accomplishments of the painters of the caves at Lascaux: brilliant work, despite limited technology. These generations may well feel that the invention of the computer marked the beginning of the next level of art, music, and literature, as it did of many other features of their civilization.

9

The Impact of Computers
on Everyday Life

The other chapters in this book have focused on the grand and global effects of computer technology, the effects on education, on science and mathematics, on civilization as a whole. But the impact on our everyday personal lives will be extraordinary and unprecedented as well.

Many of the changes wrought by the new computer technology will be nearly invisible to the user. Computers will be embedded in familiar products and appliances, and the consumer will be aware only that these products now have more capabilities, better performance and reliability, and lower cost. The magnitude of the phenomenon of "invisible" computing can be illustrated with one arresting statistic on computer chip production: The dollar value of 8-bit computer chips (such as the Z80) produced and sold in 1996 was on the order of $5.7 billion. In contrast, sales of more advanced 16- and 32-bit chips came to only $1.6 billion (*SMT Trends* 1997).

The reason this number is extraordinary may require a bit of explanation. As I write this in 1997, CPU chips have gone through at least five "generations" from the 8-bit Z80 to the 16-bit 80286 to the 32-bit 80386, 80486,

Pentium, and beyond. Indeed, the 8-bit chips are so obsolete that some 8-bit computer systems are already behind glass in museum displays. Yet 8-bit chips are still being produced in record numbers. There are three basic reasons for this: First, the 8-bit chips have computer power comparable to a mainframe computer of the early 1960s; second, the chips can be produced for less than a dollar each; and third, there is a huge demand for them. None of this vast volume of 8-bit chip production is going into conventional personal computers as it did a decade ago. Today all of it is going into appliances, microwave ovens, televisions, telephones, dishwashers, videotape recorders, and other products where the users are seldom aware that they are operating a sophisticated computer. The users know only that the capabilities of these appliances have multiplied with little change in their overall price.

Perhaps the most familiar symbol of invisible computer power is found in the humble wristwatch. A generation ago the finest wristwatches were mechanical marvels containing intricate machinery of miniaturized springs and gears that were crafted with jewel-like precision. In contrast, today's wristwatch has almost no moving parts. I am presently wearing a watch that has a quartz oscillator accurate to a few seconds per month. The watch has a calendar display that not only knows the days of week and the duration of each month but is programmed to handle and display leap-year dates correctly until the year 2100 (the next four-year span that lacks a leap year). It also has an alarm buzzer that can be set for a particular time of day and for a particular day of the year. It has a stopwatch and a count-down mode and a dual-time mode (for keeping track of two time zones) and twelve- or twenty-four-hour display. All of this technology is available at a cost of about $25.

Nothing remotely comparable was available in watches costing a hundred times as much as recently as a generation ago.

Another dramatic impact of computer technology will be found in personal transportation, particularly in the operation of automobiles. It has often been remarked that the average automobile today has more on-board computer power than the Apollo moon lander. Most of this technology is presently used to adjust and refine engine performance, and it is therefore in the "invisible-to-the-user" category. But new computer technology is being developed that will have much more visibility to the operator or driver. Howard (1997) recently described Global Positioning System (GPS) satellite-positioning technology that can be coupled with computer displays to show the current position of a vehicle on a computer-displayed map. This technology is already available and will become universal as its cost plummets. This navigational software can be programmed to give piloting instructions, by voice if desired ("Turn left at the next intersection." "You missed the turn at the last intersection. Should I recalculate the best route from this point?").

Fully automated automobile operation (driving) is almost possible with this technology, except for some minor details such as collision avoidance. And computerized technologies are being developed to handle collision avoidance as well through the use of radar, sonar, and optical sensing techniques. Such technology is already available for aircraft, and one freeway in California is already being outfitted with instruments as a test site for automatic automobile piloting. The technology will have to be made nearly fail-safe, or at least it will have to represent a substantial improvement over ordinary human operation of the automobile, but this should not be a difficult standard to achieve.

The popularity of personal finance, checkbook balancing, and tax preparation software indicates another broad area in which computers can have an impact on our personal lives. All of us have a substantial need for the tedious and extensive arithmetic operations involved in the calculation of tax payments, planning of estates, and general financial decision making. In the past there were only two options for doing this kind of work: You could hire expensive professionals to do the work for you, or you could attempt the detailed calculations yourself, slowly, tediously, and frequently with errors. Today, computer software is available that can simplify and streamline these tasks and bring fast, accurate, and inexpensive results within everyone's reach. Even minor legal matters such as writing wills and filing routine court papers can be handled by the appropriate software.

Also in the area of personal finance, a variety of banking services using home computers is becoming available. Bank customers can check balances, transfer funds between accounts, and even pay bills automatically. And sophisticated securities transactions are already being handled by computerized trading software. There is no reason why such software could not be improved to the point that it would provide trading capabilities that are more efficient, more accurate, more up-to-date, and far less costly than is possible in dealing with conventional brokers.

Another area being revolutionized by computer technology is personal communication. Cellular phone technology has changed the way we think about telephone service. This technology has led to unexpected adventures such as the case of the party of hikers in the San Juan Mountains in Colorado, one of whom fell and suffered serious injuries at a point that was many hours of hiking distance from the nearest road. Instead of hiking out, one of the other members of the party retrieved a cellular phone from his backpack, dialed 911, and telephoned out the GPS satellite coordinates of the party to a rescue crew. A helicopter evacuated the injured man within the hour.

Such communication technology has the capability to place everyone within reach of a phone call at all times. Whether this is desirable may be debated, but the popularity of cellular phones and pagers suggests that there is a substantial demand for the capability. Other technologies, such as phone answering machines and computerized voice mail, have changed the way we use telephones. Today the technology for automatic telephone answering is common and inexpensive. It allows each home to have the capability to handle telephone calls when no one is home, and it also allows for the screening of incoming calls to eliminate annoying telephone solicitations and other misuses of the telephone system.

The use of electronic technology for photography, both still and video, is growing explosively. This technology allows us to do more than record our vacations and our children at play. It may have a deep impact on law enforcement, for example, as the Rodney King case in Los Angeles illustrates. And a recent advertisement for an electronic still camera contained an unconscious irony: The advertisement touted the camera's use for police work in documenting evidence and crime scenes, and then described how a realtor could use it to photograph a property for sale and then digitally alter the image to remove unsightly trash cans. The incompatibility of the two uses was apparently lost on the advertising writer.

Computers have already had more impact on the operation of our homes than many of us are aware of, thanks to invisible computers. The simple thermostat on the wall of my own house, for example, is no longer the humble bimetal strip of years past, but rather is a sophisticated computer connected to several digital thermometers. The computerized thermostat is programmed to adjust not just the temperature of the house but the rate of change of temperature in order to achieve optimum performance of the furnace/air conditioner system at varying times of the day and week. Whereas earlier models of timed thermostats merely changed the temperature setting at a given time, this thermostat holds a target time at which the new temperature setting is to be achieved. It then checks to see if the desired time is actually met, and adjusts itself when the target is missed.

Computerized home security devices are already available. These devices can alert the police or fire department if an intruder or fire is detected. Motion sensor technology can be used to turn on lights outside the front door if someone approaches. Future advances in home security may include such things as door locks operated by fingerprint or retinal pattern recognition.

There is little need to comment on the impact of the use of computers for games and entertainment beyond the possible educational uses of such software that was discussed in chapter 5.

Another modest impact of computer technology is found at the grocery store check-out counter. The barcode scanners provide fast, efficient, and accurate checkout, and operating them requires less training than operating conventional cash registers. There is some concern about the accuracy of the programmed prices, but even this is something of an advantage over old methods of grocery check-out because the errors in the programmed prices can be checked much more easily than random clerical errors. And by creating a computerized record of each transaction, the computerized software can make the process of inventory management and restocking much simpler and less labor-intensive.

But the overall impact of computer technology on marketing and retailing will extend well beyond supermarket check-out lines, of course. For the first time, this past Christmas I did some of my own holiday shopping using the Internet to find products, compare prices and features, and place orders. The ability of computer networks to provide product information and order/sales information could revolutionize marketing. However, if cable TV shopping channels are a sample of the directions in which this technology will develop, then perhaps the less said the better.

The computer revolution also holds great promise for vastly improved efficiency in the use of resources and concomitant reduction in waste. Improved communication technology reduces the need for transportation, for example. Telecommuting is already reducing or eliminating the need to physically commute to work. And computerized optimization of travel media, including such things as improved and optimized airline schedules, vastly reduces fuel waste. Similarly, computers can improve the efficiency of freight transportation by train. In the precomputer era railroad cars were often idle because of the computational difficulties involved in optimizing their schedules. Also there are novel computer algorithms for linear programming problems that can optimize the use of resources and inventories by factories. And computerized optimization of time-critical scheduling can greatly improve the efficiency of industries such as home and office construction. The opportunities for using computer technology to reduce or eliminate waste are almost limitless.

Improving the efficiency of industry should have ripple effects throughout the entire economy. To take just one important example, the recent development of just-in-time inventory management for manufacturing has reduced the reliance on costly and inefficient inventory stocks. These techniques have the potential to ameliorate or even eliminate the classic "business cycle" of boom and bust, because those cycles were driven in some measure by excessive build-up of inventories during the "boom" portion of the cycle. If business cycles are indeed eliminated or even just reduced, the effects will be felt personally by a large fraction of the population. None of us (except perhaps economists and opposition politicians) will miss the recurrent cycles of recession and depression.

Another area that has long been touted as ripe for revolution by computer and communication technology is the publishing industry. Some newspapers and journals are already distributed electronically rather than on paper. As this technology becomes widespread these industries will become far more efficient, distributing information that is more up-to-date and, as a side bonus, becoming more environmentally friendly through a vast reduction in the use of paper. Already the use of e-mail is reducing paper usage while at the same time providing faster and more convenient communication than is possible with conventional postal services.

Yet another area that will have a broad and enormously beneficial impact on many lives involves computer aids for the handicapped. Voice recognition and speech generation technology can be of enormous benefit to the blind as well as to the illiterate, and e-mail opens many of the capabilities of telephone communication to the deaf. Quadriplegics can use detectors based on eye and mouth movements for communication. And the GPS technology described for navigating automobiles has already been adapted for use by the blind, allowing them to navigate easily and unaided to unfamiliar locations. Such aids for the handicapped may not affect many of us, but for those who are benefited the effects are of incalculable value.

Even our system of government is being changed in countless ways by computer technology, with effects ranging from computerized vote tallying to sophisticated polling techniques and targeted political advertising and fund raising. The effects are not wholly bad: The impact of fax technology and e-mail on political matters such as the Tiananmen Square protests in Beijing has been widely reported and discussed. These technologies pose both promises and dangers that will have to be dealt with in the next level of civilization.

The impact of the Internet is perhaps the most dramatic example of the ways that computer technology is transforming everyday life. "Surfing the Web" has transformed the way many of us conduct large portions of our lives, from job searching to socializing to researching important or trivial decisions. It has put vast volumes of information at our fingertips and in our homes: Without leaving my house I can search the catalogs of several nearby libraries, scan airline schedules and order tickets, check weather forecasts across the country, keep tabs on various political organizations and government agencies, and pick up information that I hadn't even guessed was in existence. The ability to "click" on keywords of Web pages opens up networks of information that can take us in completely unexpected directions.

Michael Dertouzos of MIT's Laboratory for Computer Science has a new book that itemizes in considerable detail many of the ways we can expect the information revolution to affect our daily lives (1997). The principal focus of his book is on information networks such as the Internet. Dertouzos describes a number of novel concepts including such things as "reverse advertising," the electronic equivalent of newspaper "want ads." With reverse advertising, a prospective buyer would post in an electronic marketplace a precise description of the article that he or she wishes to buy. Potential vendors would scan the marketplace and make bids. As Dertouzos notes, such a system could be advantageous to both buyers and sellers. And it could be extended so that vendors sell not just existing stock, but would actually manufacture the requested product to order. Such customized mass production or "mass individualized production" is feasible only with the enormous information-processing capabilities of computer networks (1997, 127). Other novel applications of computer networking (among the many described by Dertouzos) include such things as virtual neighborhoods, automatic house doctors, and interactive art. Dertouzos is able to encapsulate the impact of the entire computer revolution through the use of an apt metaphor:

> Information technology would alter how we
> work and play, but more important, it would
> revise deeper aspects of our lives and of humanity: how we receive health care, how children
> learn, how the elderly remain connected to society, how governments conduct their affairs....
> Most people had no idea that there was a tidal
> wave rushing toward them. (1997, 5–6)

An effective way to cultivate our understanding of the breadth of the impact of computer technology on our everyday lives today is try to think of activities, either our own or those of our neighbors, friends, and coworkers, that are *not* radically altered by computer technology either presently or in the foreseeable future. The list grows shorter by the year. At one time it seemed to me that mundane housekeeping tasks such as cleaning bathrooms might be immune to this revolution. Not so. Robotic self-cleaning bathroom technology has already been developed and will become widely available as the costs drop. The capabilities for robotic cleaning services vastly extend the areas that are impacted by computer technology. I am hard put today to come up with anything at all that is not already affected in some way, or has no potential to be affected by computer technology. The present impact of the computer revolution is already vast beyond our imagination. And this revolution has been going on for less than a generation.

It can be interesting to extend this idea into the past by trying to imagine what it must have been like to live in eras that had substantially lower levels of information production. What was it like to live before the printing press, for example, when almost everyone outside of the clergy was illiterate, when books were rare and costly works of art, and newspapers and other periodicals simply did not exist? Then imagine people in this situation trying to forecast the changes that the printing press would introduce into civilization. Could they have anticipated newspapers, for example? They would not have been able to even conceive of most of the commonplaces of our existence today. Imagining such an existence and trying to understand the differences between life in that era in contrast to the civilization that we are familiar with today can give us insight into the changes that are being wrought now by the computer revolution. It will help us to appreciate the magnitude of the changes that we can expect to see with the dawning of the next level of civilization.

10

On Growth

A book about the dawn of a new level of civilization would be remiss if it did not include at least some discussion of the overwhelming problems that the new civilization will face. The critical problems include such things as overcrowding, pollution, depletion of resources, and loss of biological diversity. Many of these problems share a common root: The population of human beings on the planet is far too large. William Rees of the University of British Columbia recently calculated that to sustain even the current population of the Earth at the average living standard enjoyed in Canada today would require two additional Earths (Moffat 1996).

But the really frightening thing about the population problem is not its present size, which is bad enough. The far more serious problem is that it is growing exponentially. Exponential growth functions are frightening even to mathematicians, who generally regard the class of problems whose difficulty grows exponentially as intractable even in theory.

It is not difficult to understand why exponential functions arouse such terror. The necessary calculations are simple enough to be within reach of a competent fourth-grader. All you need to know is how to multiply.

A pocket calculator will help; better yet is a computer spreadsheet. To calculate an exponential growth function, you simply take a starting value, multiply it by $(1 + x)$, where x is the fractional growth, and repeat the process. For example, if you want to explore an annual growth rate of 3 percent (i.e., a fractional growth rate of 0.03 per year), you multiply your starting value by 1.03, then that value by 1.03, and repeat as often as you like. The successive values give you the total quantity in each successive year. One advantage of doing this calculation with a spreadsheet is that you can instantly graph the results. Also you can change the fractional rate, and immediately the entire calculation is finished and graphed. These features are nice, but you can get by with using a pocket calculator or even (heaven forbid) pencil and paper.

The mathematically sophisticated can use the "exp" (exponential) key that is found on many pocket calculators (and the comparable function on spreadsheets) to skip all of these multiplication steps, but why would you want to miss out on all the fun?

To understand why exponential growth functions are frightening, recall the story of the king who, wishing to reward a loyal subject, asked him to name any prize he liked. The loyal subject, who was perhaps more gifted at mathematics than diplomacy, asked the king to give him a single grain of wheat on the first square of a chessboard, two grains on the second, four on the third, and so on, doubling each time through the sixty-four squares of a chessboard. According to the story, the king immediately granted the request, thinking that a few bushels of wheat was all that would be needed. However, as any computer spreadsheet or pocket calculator can determine in a few seconds, the amount of wheat requested was about 9.223×10^{18} grains, or about 598 billion metric tons.

The point of the story is that any quantity that doubles itself repeatedly will grow not only to unreasonable and even incomprehensible size; it will do so in a stunningly short time. And all exponential growth functions share this key characteristic of the wheat grains on the chessboard—they double their size at regular intervals. Determining the doubling period for any given growth rate is easy using a calculator or spreadsheet: Start with a value of 1, perform the iterated multiplications specified above, and check when the value reaches or exceeds 2. For 3 percent annual growth rate, the doubling period is about 24 years. For 1 percent, it is about 70 years. But no matter what the growth rate is, there is a period over which the growing quantity doubles.

This presents an obvious and unavoidable problem for population growth: Any nonzero exponential growth will exhaust any finite resource not only in a finite time, but in a fairly short amount of time. The existence of a finite limiting resource places a fundamental mathematical limit on growth. It does not matter what the limiting resource is, so long as it exists. Let us look at some of the more outrageous possibilities. We could assume that the limiting population of the planet is the point at which everyone occupies one square foot of land space. If we take the present population and assume a 3 percent annual rate, then this limit will be reached in less than 500 years. If we assume that the entire mass of the Earth could be converted to human biomass (using some fancy nucleosynthesis to create the right mix of atomic types), it would take only about 1,000 years to reach this limit.

Thus any reasonable limits on population size (not to mention some unreasonable limits) are going to be reached within the next millennium, assuming finite resources and an annual growth rate of about 3 percent. But an even more extraordinary feature of exponential growth rates is that they are able to overwhelm not merely finite resources but even infinite resources, and do it in finite amounts of time. To demonstrate this we could assume that the Earth is flat, that is, a plane of infinite extent containing infinite resources. Or we could assume that the human race is capable of expanding forever into intergalactic space, again with infinite resources available. (We will ignore the question of whether space itself is infinite.) Still, after a relatively short time interval, even these infinite quantities of resources will prove to be inadequate to support exponential growth.

The problem is that there is a speed limit in the universe: You cannot move faster than the speed of light. Thus even when an infinite quantity of resources is available you cannot reach those resources fast enough to support a population that is growing exponentially. Again, the calculation is simple. Starting from the present population of the planet, and assuming a 3 percent growth rate, the human biomass becomes a sphere expanding at the speed of light in about 3,300 years. Here the limiting resource is the volume of space itself. This is a limit that cannot be exceeded by any physical process.

And even in the rather unlikely event that we find out that Einstein was wrong, and exceeding the universal speed limit is possible after all, how could anyone live where the population biomass is growing at such a rate? And who would want to? The logistics are impossible even if it were possible to travel faster than the speed of light.

Clearly, this limit redefines the word absurd. It shows the fundamental reason why population problems can never be solved by colonizing outer space. And it demolishes the argument that we can rely on future technological advances to provide sufficient resources to support exponential population growth. No technological advance could possibly provide infinite resources, and even if it could, those resources would still not be adequate to support exponential growth.

Therefore exponential growth rates cause serious, unsolvable problems. Exponential growth cannot be sustained even for relatively short periods of time. But would it be possible to sustain slower (subexponential) growth rates, perhaps linear, or even slower rates such as rates that grow with the square root of time? Such rates are mathematically possible, but they all encounter three fundamental problems: First, nonzero growth of any form will eventually exhaust any finite resource. Given that all resources are finite, this by itself is an insuperable problem. Second, any growth rate that is slower than exponential will tend to zero on a per-capita basis. Only an exponential growth rate can maintain the per-capita increase rate (indeed this property virtually defines exponential growth). And third, population growth rates are naturally exponential. Thus, although subexponential growth rates are theoretically possible, in practice they are no easier to attain than a rate of zero. Attaining such a rate would require the same type of difficult moral choice that a zero rate would.

It is therefore an elementary demonstration that zero is the only growth rate that can be sustained indefinitely. We can no more sustain a growth rate larger than zero than we can force 2+2 to equal 5. Anyone who suggests otherwise is either inadequately acquainted with arithmetic or else deliberately mendacious. But nonzero growth is very similar to a narcotics habit: In the beginning it feels good, and it relieves pain. This is why it is so popular with politicians and businesspeople, among others. But eventually it must end, and the ending generally causes extreme pain (famine, war, etc.). A politician who suggests that growth can solve the problems of society should be treated like a racketeer promoting a Ponzi scheme, for that is exactly what he or she is doing.* Phrases such as "controlled growth" or "managed growth" should be recognized as oxymorons that are used only by the ignorant or the dishonest, unless they are used to mean zero growth.

We know, then, that all population growth rates must equal zero over the long term. And they will approach zero, with mathematical certainty. So why is there a problem? The problem, of course, is that every method that might be used to reduce the growth rate to zero runs headlong into a set of nearly impossible moral difficulties, because there are only two alternative ways to decrease the growth rate: We must either decrease the birth rate, or increase the death rate.

Some might think that this is a false choice, that in the long run the death rate must equal the birth rate. But "in the long run" is not good enough. To achieve zero growth the birth rate must equal the death rate simultaneously, at least on average. And today it does not. Thus we have only two options.

Increasing the death rate by slaughtering your neighbors and moving into their turf has been the classic, almost universal solution to population growth problems from the *Iliad* and the Punic Wars to Bosnia and Rwanda today. You simply divide the world into "us" and "them," and decide that it is all right to kill "them." For the discussion that follows I will assume that this is no longer permissible, that there is no morally acceptable way to increase the death rate.†

* There is no mathematical objection to exponential growth for very brief periods, but anyone who advocates growth for a finite period only should be expected to specify the period, when it will end, how it will end, and what conditions will entail when it does end.

† The argument over assisted suicide for the terminally ill is the only case I am aware of in which a moral argument for increasing the death rate is made by some today. Whichever side of this argument is right, it does not affect enough people to be relevant to questions of controlling population size.

Given this assumption, we must decrease the birth rate. Period. This is forced, with mathematical certainty. The great difficulty is that although there are many technologically feasible methods of decreasing the birth rate, none of them is morally acceptable to everyone. Thus we face a simple but vexing alternative: Are the moral objections to decreasing the birth rate less than the objections to increasing the death rate? If they are not, then we face a genuine quandary, a moral dilemma with no solution. We must select one or the other, and both are equally morally objectionable.

The only possible resolution to this dilemma is the development of a technologically feasible solution that reduces the birth rate and at the same time resolves everyone's moral difficulties. Obviously the best solution would be one that is entirely voluntary. There is some indication that this might be possible: Birth rates are down sharply in most of the developed world today.

If voluntary means prove inadequate, does it then follow that involuntary methods would be justified? Should childbearing be permitted only by government license, for example? This may seem absurd, yet we already trust governments to regulate such things as automobile and aircraft operation. Is having a child any less a responsibility than piloting an airplane? Government also regulates areas of personal conduct that have reproductive implications, such as marriage. Polygamy, for example, is outlawed despite some religious objections to the proscription. And would a government licensing program be worse than massive famine, for example? Is there any better way to achieve zero growth?

No one today has answers to these questions that will satisfy everyone, and perhaps no one ever will, but if we are to have any chance at all to resolve them they must be raised and debated. Once again, advancing technology has raised serious moral questions that simply did not exist a few generations ago, and we have no choice but to face them now. The question of population growth is perhaps the most fundamental of these questions. Gutmann and Thompson (1996) have some detailed and thoughtful discussions concerning the nature and conduct of these vital debates. They must be resolved within the bounds of mathematical possibility, and the only mathematical possibility is that the growth rate will be reduced to zero, either by reduction of the birth rate or increase of the death rate. The choice is ours.

Conclusion

Today, at the very beginning of an entirely new level of civilization, a number of writers are proclaiming the end of things. Some argue that we are at the end of physics (Weinberg 1992, Lederman 1993), and the end of science (Horgan 1996), and even at the end of history (Fukuyama 1992). Though these pronouncements may seem odd, they are a characteristic, almost requisite feature of the beginning of a civilization. As Morison notes: "At the end of the year 1492 most men in Western Europe felt exceedingly gloomy about the future" (1942, 3). Morison goes on to catalog several pages of the perceived woes of Western civilization during the Renaissance.

There is a simple reason for the gloom-and-doom projections that characterize the beginning of civilizations: Beginnings are necessarily juxtaposed with endings. Just as the invention of the printing press marked the beginning of Level 3 civilization, at the same time it marked the end of Level 2 civilization. The invention of the computer similarly marks the end of many of the familiar and characteristic features of Level 3 civilization. Level 4 civilization will be so different that many of the fondest labels from Level 3 may become obsolete. Terms such as "science" may seem as quaint to our descendants as the nineteenth-century term "natural philosophy."

My strong personal preference is to focus attention on the things that are beginning rather than on those that are ending. The analogues to physics, natural philosophy, and history that will exist in Level 4 civilization, although they will be radically different than in Level 3, will be no less exciting and interesting. On the contrary, they will be breathtaking beyond our present imagination.

The principal driving force behind the immense changes taking place in civilization today is the phenomenal development of computer technology. It is no exaggeration to say that the invention of the computer is the most important event in the history of technology, if not in history, period. The inventions of printing, writing, and language, in that order, are the next most important inventions after the computer in terms of their overall impact on civilization. But the computer is more important than these other three inventions by a rather wide margin. Computer technology is important both for the incredible breadth of its impact—it touches everything—and for the depth of the impact: Virtually everything that it touches is changed beyond recognition.

To understand both the breadth and the depth of the impact of the computer revolution in the future, try to imagine what life would be like today without its three predecessor inventions. This is a nearly impossible task because there is hardly a waking moment in which each of us is not making use of some combination of these critical inventions. And the pervasiveness of the computer revolution is already so great that there is hardly a waking *or* sleeping moment in which we are not making some use of this remarkable new technology (if only in the thermostat on the wall and the digital alarm clock).

Since this extraordinary technology has placed us at the dawn of a new level of civilization, it is only natural to wonder about the nature and characteristics of the new civilization. Unfortunately the possibilities are too numerous to allow detailed predictions with any confidence. And the history of predictions of the future of civilization holds many cautionary examples of individuals, many of them of outstanding intellects, whose attempts to forecast or even shape the future of civilization failed to match reality. Most of them did not even come close to the real world. The list begins with Plato, and includes More, Hobbes, Rousseau, Butler, Marx, and many others. Most of their forecasts were little better than amusing, and seen from the vantage of their future, many were just plain bizarre. Plato, for example, laid out an ideal civilization ruled by philosophers (what else would you expect a philosopher to propose?), supported by a warrior class that would own no private property but would hold even wives and children in common. The conspicuous absence of historical accuracy in forecasts such as these is at once cautionary and comforting: I could hardly make a worse prediction than Plato, and he is widely regarded as an individual of no mean intellect. Nevertheless, it is likely that much of what I write will serve to provide future generations with "a source of innocent merriment" (apologies to W.S. Gilbert).

Although we cannot predict all of the features of the new civilization, we can, as I noted in chapter 1, at least make conjectures about some portions of it. In order to have a strong, secure, and enduring civilization, four elements would seem to be essential: a stable population, a steady supply of energy and resources, freedom from tedious and mindless labor, and an education system that allows everyone to develop and make full use of his or her creative abilities. Other desirable aspects, such as a universal and regular language and an optimized arithmetic system, would be nice things to have, but civilization could struggle along without them. Almost anyone could make a personal wish list of additional desired qualities, ranging, say, from absence of disease to transportation systems that resemble *Star Trek* transporters.

Of the four critical elements, the first two are deeply interrelated. The population of an enduring civilization must not only have a zero growth rate, it must be small enough to live comfortably within the bounds of renewable energy sources and recyclable resources; in other words, it must be small enough to be supported by energy sources such as solar and hydroelectric power, and biomass fuels.

What is the optimal size of such a population? There is a poorly known minimum population level, probably in the range of a few tens of thousands, that is needed to preserve sufficient genetic diversity to prevent serious inbreeding. The maximum might be as low as a few tens of millions, depending on the efficiency of solar energy converters and resource recycling technology. These numbers are certainly debatable within one or two orders of magnitude. And we should not neglect the necessity of maintaining the global habitats needed for species diversity—in a word, wilderness. I would not wish to live in a world that lacked forests, tigers, orioles, or dolphins.

Element number three is the one that may mark the greatest break with past societies. Throughout history most human social organizations have contrived ways to eliminate toil for a small, privileged, and occasionally educated ruling class. Today the technologies are being developed that could allow us to bestow this privilege on everyone. It was technologically impossible to do this in the past; the ruling class had to be supported by a far larger class of servants, artisans, serfs and/or slaves. Such societies have exhibited the problems and instabilities that might be expected from a social structure that is consciously unequal and unfair: The servant class is frequently discontented with its lot, as Spartacus and Wat Tyler and Nat Turner remind us.

The possibilities of using computer and robotic technology to provide freedom from tedious labor have been discussed extensively by Moravec (1988). He argues that computer technology will have achieved human-level intelligence by the middle of the twenty-first century. I happen to think that this assessment is rather optimistic, but for the elimination of mindless labor we do not actually need robots that can match human intelligence. A common washing machine is able to eliminate such labor with almost no intelligence. Other mindless tasks such as vacuuming rugs automatically require little more intelligence. Indeed, this is the defining property of mindless tasks.

Dertouzos discusses the difficulties of what he calls the "work-free society" (1997, 271–75). He notes that the technical difficulties involved are somewhere between "improbably difficult and impossible." Perhaps no less important: "At a philosophical and psychological level, we do not have a precedent or an ideology for a work-free society." Nevertheless, he feels that such a society is an idealized limit toward which it is worth working, even if its final attainment remains perpetually out of reach.

Universal freedom from tedious labor would require substantial improvements in the technologies that are needed for the production and maintenance of the physical necessities of life—food, clothing, shelter, and so on. The Industrial Revolution made enormous strides toward reducing the need for arduous physical labor—tractors replaced horse-drawn plows, steam shovels replaced hand picks and shovels, and so on. Computer technology will allow a similar elimination of the need for tedious, repetitive, and uncreative mental labor. In a truly civilized society no one should spend a significant portion of his or her lifetime doing work that could be done by machines. In such a civilization, all labor except that which requires creative thought should be relegated to mechanical servants.

An enduring demand would exist only for the highly skilled labor that includes the ability to solve difficult and unstructured problems. Occupations requiring only manual skill and routine clerical skill are already under an inexorable squeeze as the cost of automation declines rapidly. These jobs will continue to exist only as long as their cost remains below a rapidly shrinking cost of automation. We should waste no tears over the loss of these uncreative, mindless occupations. They were never fit for human beings in the first place; eliminating them is a moral obligation. Rather than paying people to behave like robots, we should let robots behave like robots, and let people do the jobs that make use of the unique creative abilities of the human mind. As the IBM slogan expressed it: "Machines should work; people should think." Plato's goal for his ideal republic was a civilization governed by philosopher-kings; our goal should be a republic of philosopher-citizens.

In some ways a mechanical servant class would be superior to a human one. It would possess some phenomenal, superhuman capabilities: It would not require sleep, for example. The technology to achieve this goal is not quite within reach today, but there appear to be no fundamental difficulties that might impede its development. This new computerized servant class would be radically different from those of the past, and it would completely avoid the moral difficulties involved with slaveholding and other inequities through the ages, but it may not avoid all of the practical difficulties involved with the existence of a servant class. For example, training the servant class was generally a problem, and mechanical servants are no exception. The demand for computer programmers is exploding. It could be one of the great ironies of history if the task of training mechanical servants proves to be as arduous as doing the work without them. Also, as Dertouzos noted, it may not be humanly possible to actually construct *all* the necessary hardware and software to perform all possible tasks. The difference between being able to perform any given task and being able to perform all such tasks is significant. Indeed, the elimination of all mindless tasks is itself anything but a mindless task. It will require considerable genius. Yet we should be able to approach or work toward the limit that all tasks that do not require creative thought are programmed to be done by machines.

The fourth critical element (education) is deeply intertwined with the third. Freedom from tedious labor and improved education would each lose much of their value if either were developed alone. There would be little benefit in training all citizens to use their creative abilities if they had to spend a large part of their lives engaged in mindless toil. But similarly, if we should succeed in eliminating the need for toil, we will have to develop citizens who are not merely capable of creative endeavors in science, art, poetry, and other artistic fields, but are enthusiastic and eager practitioners of these skills as well. The alternative might be degeneration into decadence, characterized by idleness, crime, drug abuse, and other ills that plagued earlier civilizations, including our own, of course. Education systems will be even more essential in the future than they have been in past civilizations.

Naturally, having production and many services

taken over by machines will fundamentally alter many of the concepts that presently form the foundation of human society. Basic concepts of wealth and property will undergo significant alteration. As Arthur C. Clarke put it:

> I also believe—and hope—that politics and economics will cease to be as important in the future as they have been in the past; the time will come when most of our present controversies on these matters will seem as trivial, or as meaningless, as the theological debates in which the keenest minds of the Middle Ages dissipated their energies. Politics and economics are concerned with power and wealth, neither of which should be the primary, still less the exclusive concern of full-grown men.
> (1964, xi–xii)

Instead, one of the primary concerns of any full-grown person in a truly civilized society should be the exploration of the infinities of the marvels of this universe, the unlimited wonders of mathematics, the sciences, and the arts. Developing our full creative potential in these directions is an objective worthy of the new civilization.

This is not a full prescription for an ideal civilization, but it seems to be a minimum. It is hard to imagine an ideal civilization that lacks any of these features. The hope is rather that these elements might provide a framework within which an ideal civilization might be sought.

And although the new civilization may have many attractive features, the transition to this level of civilization will not necessarily be easy. Transitions are generally difficult, and the shift to a stable and sustainable population size is going to be particularly difficult. Also, the transition period is likely to be further marred by inequities involving disparities of access to the new technologies. Part of the solution to these inequities lies in improved education, but an ideal educational system will not be created overnight. Thus, although the long-term outlook for civilization may be very bright, the short-term prognosis is rather different: We may face a plethora of difficulties and problems. And as is frequently noted, we all have to live in the short term. We may well be facing a very difficult transition period ahead, whose spirit could be characterized by the classic Oriental curse: "May you live in interesting times."

But despite the obvious transition difficulties, one prediction about the future of civilization that can be made with some confidence is that, for whatever problems they may suffer, the inhabitants of the new civilization will not wish to trade their problems for ours. They would be no more likely to want to return to the earlier level of civilization found in the twentieth century than most of us today would want to return to a world without automobiles, aircraft, electric lights, interior plumbing, refrigerators, air conditioning, or telephones (the world into which many of our grandparents were born).

The level of civilization that is just beginning should be the envy of the ages. And from what we know today of the multiple infinities of things that exist for us to explore and discover, in mathematics and the sciences no less than in the arts, we will likely always be close to the beginning of some major development marked by an epoch-making discovery or invention. This will remain true no matter how much progress we make. We will forever be at the beginning of a new level of civilization. And this is fortunate, because a beginning is a fascinating and exciting place to be.

References

Introduction

Gamow, G. 1961. *One, Two, Three,...Infinity*. New York: Viking Press.

Newman, J.R. (ed.). 1956. *The World of Mathematics*. New York: Simon and Schuster.

Chapter 1

Berr, H. 1934. Preface to G. Weill, *Le Journal*, Paris. In the series *L'Evolution de l'Humanité*, edited by H. Berr.

Butler, J.N., and D.R. Quarrie. 1996. "Data Acquisition and Analysis in Extremely High Data Rate Experiments." *Physics Today* 49, no. 10: 50–56.

Dickens, A.G. 1966. *Reformation and Society in Sixteenth-Century Europe*. New York: Harcourt, Brace and World.

Eisenstein, E.L. 1979. *The Printing Press as an Agent of Change*. 2 vols. New York: Cambridge University Press.

Falk, D. 1984. "The Petrified Brain." *Natural History* (September): 38.

Gardner, M. 1989. *Penrose Tiles to Trapdoor Ciphers*. New York: W. H. Freeman & Co., pp. 183–204.

Gelb, I. 1963. *A Study of Writing*. Chicago: University of Chicago Press.

Gingerich, O. 1975. "Copernicus and the Impact of Printing." In *Vistas in Astronomy* 17, ed. A. Beer and K. A. Strand. Oxford, UK: Oxford University Press, pp. 201–7.

———. 1979. "The Great Copernicus Chase." *American Scholar* 49: 81–88.

Gould, S.J. 1993. *Eight Little Piggies*. New York: W. W. Norton.

McNeill, W.H. 1963. *The Rise of the West*. Chicago: University of Chicago Press.

Morison, S.E. 1942. *Admiral of the Ocean Sea*. Boston: Little, Brown & Co.

Parsons, E.A. 1952. *The Alexandrian Library*. Amsterdam: Elsevier.

Chapter 2

Ball, R. 1908. *A Treatise on Spherical Astronomy*. Cambridge, UK: Cambridge University Press, pp. 320–22.

Casti, J. 1992. *Reality Rules: II.* New York: John Wiley, pp. 287–370.

Dunham, W. 1990. *Journey Through Genius: The Great Theorems of Mathematics.* New York: John Wiley.

Gardner, M. 1983. *Wheels, Life, and Other Mathematical Amuse-ments.* New York: W. H. Freeman, pp. 214–57.

Hodges, A. 1983. *Alan Turing: the Enigma.* New York: Simon and Schuster.

Hofstadter, D. 1979. *Gödel, Escher, Bach, an Eternal Golden Braid.* New York: Random House.

Jackson, E.A. 1995. "No Provable Limits to Scientific Knowledge." *Complexity* 1: 14–17.

Kline, M. 1980. *Mathematics—The Loss of Certainty.* Oxford, UK: Oxford University Press.

Motz, L., and J.H. Weaver. 1995. *The Story of Astronomy.* New York: Plenum Press, pp. 311–12.

Nagel, E. and J.R. Newman. 1968. *Gödel's Proof.* New York: New York University Press.

North, J. 1994. *The Norton History of Astronomy.* New York: W.W. Norton.

Seife, C. 1997. "New Test Sizes Up Randomness." *Science* 276 (25 April): 532.

Stewart, I. 1992. *The Problems of Mathematics.* 2d ed. Oxford, UK: Oxford University Press, pp. 303–6.

———. 1994. "The Ultimate in Anty-Particles." *Scientific American* 271, no. 1 (July): 104–7.

Traub, J.F. and H. Wozniakowski. 1994. "Breaking Intractability." *Scientific American* 270, no. 1 (January): 104–7.

Chapter 3

Briggs, J. 1992. *Fractals.* New York: Simon and Schuster, pp. 73–82.

Casti, J. 1997. "Creation in Silicon." *Nature* 383: 399.

Dewdney, A.K. 1985. "Computer Recreations." *Scientific American* (August): 16–24.

Dunham, William. 1990. *Journey Through Genius: The Great Theorems of Mathematics.* New York: John Wiley.

Gardner, M. 1983. *Wheels, Life, and Other Mathematical Amuse-ments.* New York: W.H. Freeman.

Horgan, J. 1996. *The End of Science, Facing the End of Science in the Twilight of the Scientific Age.* Reading, Mass.: Addison-Wesley.

Lederman, L.M., with D. Teresi. 1993. *The God Particle: If the Universe Is the Answer, What Is the Question?* Boston: Houghton Mifflin.

Pavelle, R., M. Rothstein, and J. Fitch. 1981. "Computer Algebra." *Scientific American* (December): 136–53.

Press, W.H., B.P. Flannery, S.A. Teukolsky, and W.T. Vetterling. 1992. *Numerical Recipes.* Cambridge, UK: Cambridge University Press.

Rawlins, G.J.E. (ed.) 1991. *Foundations of Genetic Algorithms.* San Mateo, Calif.: Morgan Kauffman.

Stewart, I. 1992. *The Problems of Mathematics.* 2d ed. Oxford, UK: Oxford University Press.

————. 1996. *From Here to Infinity.* Oxford, UK: Oxford University Press.

Weinberg, S. 1992. *Dreams of a Final Theory.* New York: Pantheon Books.

Chapter 4

Courant, R., and H. Robbins. 1996. *What Is Mathematics?* Oxford, UK: Oxford University Press.

Dunham, W. 1990. *Journey Through Genius: The Great Theorems of Mathematics.* New York: John Wiley.

Gamow, G. 1961. *One, Two, Three,...Infinity.* New York: Viking Press.

Hahn, H. 1956. "Infinity." In *The World of Mathematics.* Vol. 3, ed. J. R. Newman. New York: Simon and Schuster, pp. 1593–1618.

Kline, M. 1980. *Mathematics—The Loss of Certainty.* Oxford, UK: Oxford University Press.

Newman J.R. 1956. "The Rhind Papyrus." In *The World of Mathematics.* Vol. 3, ed. J.R. Newman. New York: Simon and Schuster, pp. 170–79.

Stewart, I. 1996. *From Here to Infinity.* New York: Oxford University Press.

Turing, A. M. 1965. "On Computable Numbers, with an Application to the Entscheidungs Problem." In *The Undecidable,* ed. M. Davis. Hewlett, N.Y.: Raven Press, pp. 115–54.

Chapter 5

Dertouzos, M. 1997. *What Will Be.* New York: HarperEdge.

Kay, A. 1991. "Computers, Networks and Education." *Scientific American* 262, no. 3 (September): 138–48.

Papert, S. 1980. *Mindstorms.* New York: Basic Books.

Plato. 1901. *The Republic of Plato.* Rev. ed., tr. B. Jowett. New York: Willey Book Co.

Stewart, I. 1996. *From Here to Infinity.* New York: Oxford University Press.

Chapter 6

Aeschylus. 1956. *Prometheus Bound.* Tr. D. Grene. In *Aeschylus II,* tr. D. Grene and S.G. Bernardete. The Complete Greek Tragedies. Chicago: University of Chicago Press.

Hamilton, E. 1937. *Three Greek Plays.* New York: W.W. Norton, pp. 9-16.

Johnson, J.E. (ed.) 1944. *Basic English.* The Reference Shelf, Vol. 17, no. 1. New York: H.W. Wilson.

Pinker, S. 1994. *The Language Instinct.* New York: William Morrow & Co.

Strunk, W., Jr., and E.B. White. 1979. *The Elements of Style.* 3d ed. New York: Macmillan.

Chapter 7

Glaser, A. 1981. *History of Binary and Other Nondecimal Numer-ation.* Los Angeles: Tomash Publishers.

Knuth, D. 1981. Seminumerical Algorithms. *The Art of Computer Programmin.* Vol. 2. Reading, Mass.: Addison-Wesley, pp. 178–93.

Chapter 8

The Norton Anthology of English Literature. 1979. Vol. 2, 4th ed., ed. M.H. Abrams. New York: W. W. Norton & Co.

Schroeder, M. 1991. *Fractals, Chaos and Power Laws.* New York: W.H. Freeman.

Chapter 9

Dertouzos, M.L. 1997. *What Will Be.* New York: HarperEdge.

Howard, B. 1997. "Hardware Must-Haves." *PC Magazine* 16, no. 6 (March 25): 95.

SMT Trends. 1997. Vol. 14, no. 2 (February).

Chapter Ten

Gutmann, A., and D. Thompson. 1996. *Democracy and Dis-agreement.* Cambridge, Mass.: Belknap Press.

Moffat, A.S. 1996. "Ecologists Look at the Big Picture." *Science* 273, no. 13 (September): 1490.

Conclusion

Clarke, A.C. 1964. *Profiles of the Future.* New York: Bantam Books.

Dertouzos, M. 1997. *What Will Be.* New York: HarperEdge.

Fukuyama, F. 1992. *The End of History and the Last Man.* New York: The Free Press.

Horgan, J. 1996. *The End of Science: Facing the End of Science in the Twilight of the Scientific Age.* Reading, Mass.: Addison-Wesley.

Lederman, L.M., with D. Teresi. 1993. *The God Particle: If the Universe Is the Answer, What Is the Question?* Boston: Houghton Mifflin.

Moravec, H.P. 1988. *Mind Children: The Future of Robot and Human Intelligence.* Cambridge, Mass.: Harvard University Press.

Morison, S.E. 1942. *Admiral of the Ocean Sea.* Boston: Little, Brown & Co.

Weinberg, S. 1992. *Dreams of a Final Theory.* New York: Pantheon Books.

Index

abstract theory, 4, 5, 39
accelerator, 22, 59
adjectives, 125
adverbs, 125
advertisements, 22
Aeschylus, 128
age brackets, 124-25
age of information, 8
agreement, subject/verb, 123
algebra, 41, 42, 48, 60, 66, 72, 75-78, 80,
    84, 98, 100, 107, 108
algorithm, 50-53, 60, 62, 63, 85, 110,
    151, 166
alphabet, 20, 28, 91, 122
analysis, 48, 49, 58-61, 62, 63, 64, 69, 98,
    104, 107, 115, 134, 157
ancient times, 28, 29, 124, 129, 141
angles, geometrical, 58
animation, 152
anthropology, 11-12, 13
Appel, Kenneth, 65, 66
appliances, 160, 161
Arabic numeration, 132
Archimedes, 70
architecture, 10, 25
Argonne National Laboratory, 66
Aristotle, 30, 41, 58, 86, 144
arithmetic, 6, 26, 43, 72, 73, 76, 103, 108-
    10, 129, 131, 132, 133-42, 163, 175,
    182
Arnauld, Antoine, 76
art, 6, 9, 10, 14, 71, 94, 101, 111, 143, 144,
    146, 147, 150-52, 153, 155, 157, 158,
    159, 169, 170, 182, 185, 186, 187
astronomy, 14, 15, 17, 18, 39, 40, 55, 56,
    57, 58, 59, 61
athletics. 157
aural simulation, 146, 152
authors, 150, 154
automation, 52, 55, 63, 184
automobile, 10, 26, 29, 33-34, 162, 167,
    177, 187
axiom, 4, 41-44, 54

Bacon, Roger 30, 32, 33, 35
banking, 33, 163
bases, mathematical, 133-42
binary system, 19, 131, 132-141
biology, 32, 59, 62, 171
biomass 173, 174, 182
bit, 19-24, 26, 48, 62, 73, 133, 138, 160,
    161
blueprints, 26
bombs, atomic, 35
Boole, George, 41, 66
Brahe, Tycho, 18, 57-59
Bronze Age, 29
business, 3, 8, 98, 167
byte, 138

cable, 166
calculation, 27, 50, 53, 54, 61, 75, 109,
    111, 134, 141, 163, 172, 174
calculator, 108, 110, 111, 172, 173
calculus, 5, 57-60, 62, 64, 69, 70, 83
calendar, 161

Cantor, Georg, 3, 4, 62, 67, 72, 75, 77-84, 86, 88-92
cathode-ray tube. *See* CRT
cellular automaton, 52, 55, 63, 163, 164
central processing unit. *See* CPU
chaos theory, 38, 46
chess, 157, 172, 173
children, 34, 96, 110, 123, 164, 169, 181
chip, computer, 21, 160, 161
choreography, 151, 157
ciphers, 34
circle, 49, 58, 59
circuits, 135
civilization, 3, 5-7, 9-14, 15, 19, 20, 22-33, 35, 37, 39, 58, 70, 71, 93, 94, 112, 114, 131, 141, 143, 159, 160, 168, 170, 171, 179-182, 184, 185, 186, 187
Clarke, Arthur D., 29, 30, 186
classification system, 29, 30
codes, 34, 35, 137, 158, 165
combinatorics, 60
communication, 10, 17, 27, 114, 115, 129, 138, 163, 164, 166, 167
communications 13, 106, 107
computable numbers, 75, 84
computer, mainframe, 161
computers, 3, 5-9, 11, 20, 23, 33-35, 37, 38, 48, 50, 52-54, 56-64, 60, 66, 67-72, 84, 85, 93, 95, 96-98, 100, 102-20, 129, 131-38, 141, 143-152, 155-170, 172, 179, 180, 183, 184, 185
conjugation of verbs, 121
Conway, John, 52, 63
Cook, James, 39, 40
Copernicus, 5, 7, 14, 15, 18, 37-39, 40, 44, 46, 55, 57
correspondence, 5, 79-82, 89-91
CPU, 160
crime, 164, 185
CRT, 144, 145
crystal, 145

da Vinci, Leonardo, 14, 132, 158
dance, 146, 151, 157
Dark Ages, 28
data, 18, 22, 58-60

databases, 59
decidability, 42, 43, 50, 51
decimal arithmetic, 47, 50, 53, 76, 131-141
denumerable sets, 81
diagonal proof, 75, 89, 91
diagrams, 26
dictionary, 107, 155, 156
digits, 19, 54, 61, 84, 89, 90, 132, 137
disk, compact, 143
divisibility, mathematical, 140
DNA, 59
doubling, 20, 173
drama, 127, 128
drawing, 144
driving, 162, 180
dual case (grammar), 122, 123, 124
duodecimal system (mathematics), 133, 134

Earth, 38, 39, 44, 73, 128, 171, 173, 174
economics, 167, 186
Edison, Thomas Alva, 10
editing, 157
education, 6, 25, 93-97, 99, 100, 105, 106, 108, 109, 111, 112, 113, 114, 143, 160, 181, 185, 187
Einstein, Albert, 46, 174
electronics, 5, 23-25, 33, 34, 50, 133-35, 143, 144, 164, 168, 169
elements, 79, 82, 91, 92
ellipse, 47, 59
e-mail, 168
energy, 22, 44, 156, 181, 182
engineering 3, 4, 29, 34, 59, 94, 98, 103, 114
engines, 25, 162
English, 6, 60, 115-127, 132, 154
ephemeris, 47, 48
equation, 5, 6, 44, 48-49, 55, 74, 87, 88
errors, 17, 47, 63, 102, 163, 165
Esperanto, 115
Euclid, 40, 41, 57-59, 69, 70, 73, 78, 85, 86, 88
Euler, Leonhard, 61, 64, 65, 74, 83
Euripides, 127

examinations, 100-104
experimentation, 17, 22, 29, 44, 45, 64, 144, 146
exploration, 7, 14, 17, 25, 46, 186
exponent, 137
exponential functions, 171-75

factor, common, 87, 88
factorization, 88
fax, 168
feedback, 101-3
Fermat, Pierre de, 51, 65, 70
fingerprint, 165
forecasts, 7, 32, 33, 34, 168, 170, 181
four-color theorem, 65
Fourier, Joseph, 61, 69
fractals, 49, 55, 60, 62, 64
fractions, 76, 109, 134, 137
frequency, 54, 60, 102
Frye, Roger, 64
future, 6, 7, 33, 54-56, 65, 68, 69, 108, 113, 115, 121, 127, 129, 131, 146, 147, 150, 154, 157, 159, 165, 169, 174, 179, 180, 181, 185-87

galaxy, 39
Galileo, 78-80
games, 97-100, 104-6, 146, 157, 165
genetics, 34, 182
genome 32
geographers 16
geography, 15, 16, 97
geometry, 40, 41, 43, 45, 57, 58, 98
geophysics, 59
gerunds, 125
Gödel, Kurt, 4-5, 38, 43, 44, 48, 50, 51, 53, 72, 77
government, 8, 22, 25, 26, 33, 168-69, 177
GPS 162, 163, 167
grammar, 97, 107, 119, 125, 155
graphics, computer, 144
grouping, of bits, 138

handwriting, 16, 108, 109
hardware, 24, 117, 155, 185

health care, 169
hexadecimal system (mathematics), 137-142
Hilbert, David, 4, 40, 42, 43, 50, 51, 72
Hippocrates, 58
history, 3-5, 7-11, 16, 19, 31, 38, 40, 55, 57, 64, 75, 93, 94, 97, 100, 132, 135, 179-82, 185
holography, 145
housekeeping, computerized, 169-70
hypertext, 156-57

IBM, 184
illiteracy, 32, 167, 170
imagination, 32, 170, 180
indexes, 23
industry 17, 27, 106, 141, 166, 167, 183
infinitive (grammar), 121
infinity, 3, 4, 48, 49, 77, 78, 81, 82-84, 186, 187
inflection (grammar), 122-25
information, 5, 8-11, 13, 15-28, 30-33, 44, 46, 47, 58, 93, 99, 108, 122-26, 155, 166-70
instruction, 97-100, 103, 104, 106, 107, 109, 111
instruments, musical, 147-150, 162
integer, 78, 80, 87-89, 133
integrals, 62, 63
intelligence, 16, 23, 94, 112, 132, 183
Internet, 24, 155, 166, 168
invention(s), 3, 5, 8, 9, 11-15, 17, 18, 31, 34, 37, 53, 57, 58, 61, 64, 70, 72, 75, 76, 108, 110-112, 133, 134, 143, 152, 159, 179, 180, 187
inventories, 166, 167

Julia, Gaston, 62

Kepler, Johannes, 57, 59
keyboard, 108, 117, 149
keys, 148
keystrokes, 150
knowledge,11, 21, 28, 38, 39, 45, 53, 56, 58, 96-98, 100, 102, 104, 107
Koch, Helge von, 62

language, 5, 6, 8, 9, 11-13, 20, 21, 24, 114-22, 124-30, 132, 143, 151, 180, 181
laser, 45, 147
law, 9, 10, 17, 30, 35, 54, 55, 141, 164
learning, 97, 98, 101, 107-10, 115, 120-22
letters (alphabetical), 20, 23, 108, 127, 139, 150
libraries, 15, 21-24, 26, 31, 109, 168
life, 22, 29, 52, 53, 55, 61, 63, 93, 98, 100, 154, 160, 168, 170, 180, 183
limits/limitations, 5, 9, 10, 19, 20, 23-26, 27, 28, 30, 31, 38, 46, 47, 50, 52, 53, 54, 59, 61, 63, 71, 74, 75, 77, 84, 133, 144, 145, 148, 149, 154, 155, 159, 173, 174, 183, 185
literacy, 25
literature, 9, 10, 13, 25, 60, 81, 94, 127-29, 152-59
logarithms, 61, 74, 135
logic, 41, 42, 58, 97
LOGO, 98, 107, 110
Luther, Martin, 14-16

machine, 5, 23, 50, 52, 53, 183
magnitude, orders of, 20-25, 38-40, 53, 54, 56, 62, 67, 73, 96, 160, 170, 182
mail, 34, 164, 167, 168
Mandelbrot set, 49, 62
manufacturing, 8, 27, 146, 147, 166, 169
mathematicians 4, 40, 41, 50, 51, 53, 55, 59, 62-71, 76, 77, 80, 81, 83, 110, 133, 171
mathematics 3-6, 14, 25, 37, 38, 40-46, 48-58, 60, 64-67, 69, 70-75, 77, 83-85, 94, 97, 103, 105, 107, 109, 112, 153, 160, 172, 186, 187
McCune, William, 66
measurement, 19, 22
memorization, 20, 109, 111, 124, 136, 139, 140
methods, scientific, 17, 18, 20, 49, 59, 61, 62, 63, 64, 67, 69, 71, 77, 90, 93, 109, 112, 115, 116, 137, 151, 165, 176, 177
metric system, 142, 172

microchip, 94
Middle Ages, 132, 186
MIDI, 148-50
millennium, 12, 117, 173
mind, 7, 14, 16, 20, 21, 96, 184
MIT (Massachusetts Institute of Technology), 4, 98, 168
music, 10, 143, 144, 147, 149-151, 159
Musical Instrument Digital Interface. See MIDI

navigation, 162, 167
network/networking, 27, 149, 166, 168, 169
Neumann, John von, 135, 153
newspapers, 22, 167, 168, 170
Newton, Isaac, 57-59, 70
Newtonian theory, 5, 47, 49, 55, 59, 70, 99
nineteenth century, 39, 41, 117, 180
noise, 149
non-Euclidean geometrics, 41, 45
nonlinear problems, 46-50
notations, 76
nouns, 122-125
nucleosynthesis, 173
numbers, 6, 10, 14, 15, 20, 21, 27, 28, 35, 39, 40, 43, 44, 46, 47, 48, 50, 51, 54, 55, 59, 61, 62, 66, 67, 71-90, 94, 102, 105, 110, 111, 112, 114, 115, 118, 120, 121, 123, 125, 132-41, 139, 140, 156, 157, 158, 160, 161, 168, 179, 182
numbers, finite, 40, 43, 44, 50, 55, 60, 66, 78, 79, 82, 85, 86, 88, 137, 173-75
numbers, hyperreal, 93
numbers, infinite, 43-47, 52, 54, 55, 73, 76, 78-81, 90, 91
numbers, irrational, 40, 47, 72-76, 78, 76, 81, 84, 88, 133
numbers, transfinite, 78, 80, 83-84
numbers, uncomputable, 6, 71, 72, 74, 75, 77, 78, 84, 85
numerals, 109, 132, 135
numeration, 132

observation, 17, 18, 22, 39, 58

octal system, 135, 137-142
office, 24, 26, 166
Ogden, C. K., 118
opera, 146, 158
operator, 162
optical sensing techniques, 162
optimization, 48, 63, 166
orbits, 49, 59, 99
organization, 35, 107, 168, 182
orthography, 117, 118
oscillator, 161
overhead, 26
overseer, 104
oversimplification, 27
oxymorons, 176

pagers, 164
paper, 48, 108-10, 145, 150, 167, 172
Papert, Seymour, 98, 107, 108, 110
paradox, 66, 72
parallax, 39
parallels, 3, 5, 7, 8, 29, 37, 41, 46, 47, 57, 58, 69, 151
parameters, 47, 61
participles, 125
particle, 59
Pascal, Blaise, 133
Peano axioms, 43
pencil, 108-10, 172
penmanship, 108
Pentium chip, 160
performers, 147-150
Pericles, 10
period, 9, 16, 28, 37, 47, 117, 118, 126, 173, 177, 180, 186, 187
periodicals, 22, 170
periodicity, 47
permanence of records, 158
pestilence, 32
petroleum, 34
philosophy, 9, 10, 19, 38, 40, 44, 94, 180, 181, 184
phonemes, 151
phonograph, 10
phosphor, 145
photography, 164

physicists, 45, 55
pianists, 147, 148, 151
piano, 148-50
pictographs, 122
piloting, 162, 177
piston, 26
pitch, 149
pixel, 145
plagiarism, 158
plagues, 32
planes, 49, 65, 174
planets, 18
Plato, 96, 111, 181, 184
play, 96, 97, 115, 128, 151, 164, 169
Pliny, 16, 17
plows, 183
plumbing, 187
plural, formation of, 122-24
poetry, 20, 25, 127, 128, 154, 185
Poincare, Henri, 62
politics, 10, 168, 186
polling, 168
pollution, 6, 33, 34, 171
Polo, Marco, 16, 17
polygamy, 177
polygon, 49
polynomials, 48, 49, 75
possessive form (grammar), 122, 124
postal system, 136, 167
poverty, 32
predictions, 32, 35, 36, 181, 187
prepositions, 121, 122, 125
preservation, 157
press, 5, 15, 16, 25, 27, 33, 57, 63, 170, 179
primes, 73, 81, 85, 86
printing, 5, 8, 9, 11, 13-18, 20, 22, 24, 25, 27, 31, 33, 57, 67, 89, 108, 113, 147, 155, 170, 179, 180
privacy, 34
problems, 6, 25, 25, 32-35, 39, 42, 43, 45, 46-51, 53, 54, 55, 59, 60, 62, 63, 65-67, 70, 76, 77, 79, 87, 88, 94-99, 101-4, 106, 109, 111, 112, 114-18, 120, 127, 129, 136, 137, 144, 145, 146, 151, 152, 156, 157, 166, 171,

problems (continued), 173-176, 183, 184, 185, 187
processor, 22, 107, 110, 111, 149, 154
production, 8-12, 16, 18-20, 27, 31, 107, 146, 149, 150, 152, 155, 160, 161, 169, 170, 183, 185
products, 160, 161, 166
prognosis, 31, 32, 187
programmers, 136-138, 141, 185
progress, 9, 28, 55-57, 62, 71, 96, 97, 100, 102, 133, 142, 147, 187
projection, 7, 146, 179
projects, 19
pronouns, 122, 124
proofs, 5, 6, 42, 51, 64-69, 73, 75, 81, 82, 85-91
proposition, 42, 43, 76, 88
protests, 168
Ptolemy, 17, 18
publications, 15, 22
publishing, 22, 167
pyramid, 29, 30, 111
Pythagoras, 88
Pythagorean theory, 40, 55, 58, 70, 72, 73, 75, 86

quadriplegics, 167
quartal system, 138-142
quartz, 161
quotients, 108

radar, 162
radio, 10
radix, 133, 134, 138
rational numbers, 72-76, 81, 86, 88, 134
reality, 145, 146, 151, 157, 159, 181
recorders, 161
redundancy, 21
Rees, William, 171
refining, 18
reform, 15, 120
Reformation, 14, 15
refrigerators, 29, 187
relativity, 45, 46
relic, 122, 124, 132
religion, 9, 10, 29, 127

Renaissance, 7, 9, 14, 16, 28, 37, 57, 93, 179
republic, 96, 184
research, 168
restraint, 33
restrictions, 9, 10, 27, 45
results, 59, 62-64, 68, 81, 101, 103, 115, 131, 134, 163, 172
retailing, 166
revision, 107, 154
revolution, 3, 5-9, 14, 18, 27, 28, 35, 37, 38, 40, 44, 46, 55, 56-62, 69, 70-72, 93, 108, 114, 119, 131, 143, 157, 163, 166-170, 180, 183
reward, 99, 102, 104, 172
Reynolds numbers, 48
robots, 170, 183, 184
rocket, 25
rules, 17, 51, 52, 55, 63, 71, 75-78, 109, 120, 122

satellites, 162, 163
scanners, 165, 168, 169
schedules, 27, 166, 168
schools, 8, 71, 94, 110
science, 3, 8, 10, 29, 40, 45, 56, 57, 59, 60, 62, 63, 64, 69, 70, 94, 103, 112, 114, 160, 168, 179, 185, 186, 187
screen (computer), 144, 145, 156
sculpture, 145, 146
securities, 163
sensing techniques, 162
sensors, 146
sentence, 122
servants, 183-185
Shakespeare, William, 60, 127, 153, 154, 156, 158
Shanks, William, 54
Shannon, Claude, 11, 19
shelter, 183
ships, 10
simplification, 121, 139, 140
simulations, 63, 98, 146
singing, 149
slaves, 183, 184
smuggling, 35

socializing, 168

society, 9, 21, 28, 31, 94, 110, 112, 133, 169, 176, 183, 184, 186

sociology, 11

software, 24, 97, 99, 104-8, 111, 112, 117, 141, 162, 163, 165, 185

solutions, 34, 46-49, 51, 59, 63, 70, 74, 95, 96, 98, 104, 116, 176, 177, 187

solvable problems, 46, 48, 49, 60

sonar, 162

sounds, 149, 150

sources, 122, 126, 182

space, 45, 56, 63, 122, 173, 174

spacecraft, 32

speech, 149, 167

spelling, 107, 116-18, 120, 130, 151, 155

sports, 157

spreadsheet, 108, 110, 111, 172, 173

stars, 39, 53, 157

Stewart, Ian, 50-53, 64-67, 69, 74, 76, 83, 109, 110

students, 70, 71, 95-101, 103-12, 126, 141

subjective case, 122

subjunctive case, 121

submarines, 32, 33

subsets, 78, 81, 82, 91, 92, 118, 120

sums, 108

supercomputer, 61

supermarket, 166

supernovae, 64

superset, 126

symbols, 161

symmetry, 46, 47, 49, 154

synthesis, 149

synthesizer, 149

synthespian, 152, 157

systems, 27, 32, 97, 113, 118, 132, 138, 139, 142, 161, 182, 185

tables, 18, 19, 63, 80, 109, 135, 139, 140

tablets, 109, 157

Tarski, Alfred, 66

tax, 162, 163

teaching, 96, 98, 103, 106, 107, 110, 151

techniques, 17, 18, 35, 45, 49, 59-63, 65, 66, 69, 71, 72, 76, 77, 99, 100, 103, 105, 107-10, 121, 137, 143, 145-47, 152, 156, 157, 158, 162, 167, 168

technology, 3, 6-10, 12, 23, 30, 31, 33-35, 37, 54, 59, 60, 63, 64, 67, 71, 93, 95, 96, 100, 102, 108-10, 111, 112-19, 130, 143-47, 148, 149-57, 159-70, 177, 180, 182, 183, 184, 187

telecommuting, 166

telegraph, 10, 25

telephone, 10, 25, 29, 161, 163, 164, 167, 187

telescope, 58

television, 144, 157, 161, 166

temperature, 165

tenses (grammar), 121, 127

teraflop, 61

tests, 101

theology, 45

theorem, 4, 38, 41, 43, 45, 48, 50, 51, 53, 58, 65

theory, 6, 10, 11, 19, 20, 34, 37, 38, 44-46, 48, 51, 55, 56, 59, 62, 64, 69, 72, 81, 83, 85, 171

thermometers, 165

thermostat, 164, 165, 180

theory of everything. See TOE

thesaurus, 107, 155

titles, 14, 22

TOE, 44-46, 50, 52-54, 56

tradition, 21

trainer, 104

transactions, 34, 163

transfinites, 80, 83

transformation, 9, 37, 61, 93

transit, 39

transition, 28, 126, 186, 187

translations, 128, 129

transportation, 27, 130, 162, 166, 182

transporters, 182

triangle, 58

truth, 5, 40-42, 153

Turing, Alan, 4, 5, 38, 43, 45, 50-53, 60, 72, 74, 75, 77, 83

Turner, Nat, 183

Tyler, Wat, 183

typesetting, 22
typewriter, 108, 110, 111, 147, 153
typist, 117
tyranny, 128

unanswerable questions, 38, 53
units, 132
universe, 7, 17, 37-40, 44-46, 50, 52-56,
    174, 186
unprovable axioms, 41, 43
unsolvable problems, 49, 60, 175
urbanization, 12
utilities, 24

verbs, 123, 125
vertices, 49
video, 164
virtual reality, 145, 146, 151, 157, 159,
    169
visibility, 162
vocabulary, 97, 118-20, 125
voice, 108, 117, 118, 130, 149-53, 162,
164, 167
voltage, 35
volumes, 15, 21, 145, 168

war, 12, 32-35, 152, 175, 176
waste, 34, 97, 114, 115, 166, 184
watches, 161
weather, 168
web, 168
wills, 163
words, 9, 12, 18, 22, 23, 26, 28, 31, 38, 40,
    42, 44, 47, 55, 60, 65, 73, 74, 81, 86,
    87, 96, 107, 108, 110, 111, 117, 119,
    121, 122, 124, 125, 127, 130, 137,
    138, 149, 152, 153, 154-156, 174, 182
writers, 6, 29, 127, 153, 155, 164, 179
writing, 5, 8, 9, 11-13, 20, 21, 24, 28, 51,
    107, 108, 110, 111, 113, 153, 155,
    157, 163, 180

zero, 89, 109, 175-78, 182